"A book on Edmund Zavitz is long overdue. Dr. Zavitz was a major contributor to forestry in Ontario in the fields of reforestation, fire prevention, and forest administration generally. I am honoured to have known him quite well and delighted to learn more about him in these pages."

— **Jim Coats, R.P.F., Former Executive Vice-President, Ontario Forestry Association**

"Edmund Zavitz has been the unsung hero of Ontario's forests for long enough. Without Zavitz, Ontario might still be the barren wastelands and blow sands of the early 1900s. One man can, and did make a difference."

— **Brenlee Robinson, Master of Forest Conservation, Vice-President of the Ontario Urban Forest Council**

Two Billion Trees and Counting

— The Legacy of Edmund Zavitz —

John Bacher

DUNDURN
NATURAL HERITAGE
TORONTO

Copyright © John Bacher, 2011

All rights reserved. No part of this publication may be reproduced, stored in a retrieval system, or transmitted in any form or by any means, electronic, mechanical, photocopying, recording, or otherwise (except for brief passages for purposes of review) without the prior permission of Dundurn Press. Permission to photocopy should be requested from Access Copyright.

Editor: Jane Gibson
Copy Editor: Matt Baker
Design: Jesse Hooper
Printer: Webcom

Library and Archives Canada Cataloguing in Publication

Bacher, John C. (John Christopher), 1954-
  Two billion trees and counting : the legacy of Edmund Zavitz / by John Bacher.

Includes bibliographical references and index.
Issued also in electronic formats.
ISBN 978-1-4597-0111-3

  1. Zavitz, Edmund, 1875-1968. 2. Foresters--Ontario--Biography. 3. Conservationists--Ontario--Biography. 4. Reforestation--Ontario--History. 5. Forest conservation--Ontario--History. I. Title.

SD129.Z38B34 2011   634.9092   C2011-902571-X

1 2 3 4 5   15 14 13 12 11

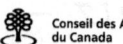  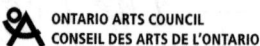

We acknowledge the support of the **Canada Council for the Arts** and the **Ontario Arts Council** for our publishing program. We also acknowledge the financial support of the **Government of Canada** through the **Canada Book Fund** and **Livres Canada Books**, and the **Government of Ontario** through the **Ontario Book Publishing Tax Credit** and the **Ontario Media Development Corporation**.

Care has been taken to trace the ownership of copyright material used in this book. The author and the publisher welcome any information enabling them to rectify any references or credits in subsequent editions.

                                                                                                                                                    J. Kirk Howard, President

Front Cover:
(Bottom) Norfolk wasteland, planting in 1912 at Norfolk Station No.1. Photo by E.J. Zavitz. *Courtesy of Ed Borczon.*
Back Cover:
(Top) Graeme Davis, county forester, at work in the Hendrie Forest as it looks today. Photo by Mark Wanzel Photography. *Courtesy of the County of Simcoe.*
(Bottom) A young Edmund Zavitz, circa 1905, taken shortly after he began lecturing at the Ontario Agricultural College in Guelph. *Archival and Special Collections, University of Guelph Library.*

Printed and bound in Canada.
www.dundurn.com

Dundurn	Gazelle Book Services Limited	Dundurn
3 Church Street, Suite 500	White Cross Mills	2250 Military Road
Toronto, Ontario, Canada	High Town, Lancaster, England	Tonawanda, NY
M5E 1M2	LA1 4XS	U.S.A. 14150

To my parents, Winfred and Mary Bacher

Contents

Foreword by Ken Armson, R.P.F.		9
Acknowledgements		13
Introduction		17
One	Edmund Zavitz: The Man Who Did Plant Trees	21
Two	Early Influences	29
Three	Behind the Scenes	49
Four	Exiled to Agriculture, 1905–1911	65
Five	The Struggle Against Indifference	95
Six	Drury and Zavitz: A Partnership	121
Seven	A Decade of Environmental Reform	145
Eight	From Disaster to Triumph	171
Nine	Implementing the Vision	197
Appendix:	Chronology of the Life of Edmund Zavitz	221
Notes		227
Bibliography		257
Index		267
About the Author		275

Foreword

CANADIANS TAKE THEIR FORESTS for granted, and the people of Ontario are no exception. The multiple packaged products that come from these forests — including the more readily identified pieces of lumber — regularly flow into our largely urban marketplace, but most individuals have little concept of the people, processes, or even the actual forests that make these commodities available. From time to time, forests command media attention and make their way into the urban dwellers' electronic viewscape as catastrophic fires, insect predations, or windstorm destruction. There are, however, small segments of the population with an interest in protecting our forests from harvesters of wood or animals. Thus, the history of Ontario's forests has been told almost exclusively in terms of either mere exploitation for furs and fibre or its presence as a barrier to settlement.

The earliest accounts of Ontario's forests were written by explorers such as Champlain, religious missionaries, traders, as well as military officers and their wives. Later, as the early waves of European settlers arrived in Southern Ontario, they, too, wrote about the forests they encountered. These accounts took the form of letters or journals, and the heavily wooded terrain appeared in the context of the settlers' everyday lives. In order to establish themselves, these settlers had to farm, and the forest had to be removed if the soil was to be tilled. To British settlers in the nineteenth century, especially those from industrial cities, these forests of large pines and hardwoods were primeval. They had no way of knowing that, especially in Southern Ontario, often the pines and other species were invaders on abandoned aboriginal agricultural lands, or had resulted from other major natural disturbances such as wind and fire.

The pines and their importance to Britain during the Napoleonic Wars were at the heart of the first major exploitation of Ontario's forests. From this came the colourful legends and myths of loggers, river drives, and the whirring, clanging sawmills around which small towns arose. The names of the major players — Booth, Dodge, and Gilmour, to cite just three — involved in this period of Ontario's history are now, if not forgotten, only recognized in the areas in which they operated. By the late nineteenth century, these "timber barons," a few farmers, and some government officials were beginning to express major concerns for both the protection of forests from fire and the reforestation of not only the timber cutovers but also the eroded and despoiled farmlands. In Southern Ontario, particularly where forests had been cleared from sandy soil — more often than not of pine stands — the erosion by water and wind had become a significant issue.

Enter into the picture a young graduate forester, Edmund John Zavitz, who was appointed a lecturer in forestry at the Ontario Agricultural College in 1905 to, in his words, "give lectures in forestry, develop forest nurseries to supply trees for reforesting, attend Farmers' Institutes and begin a survey of the large waste areas in Southern Ontario." One of the results of his report published in 1908 was the establishment of the St. Williams Forest Station and nursery. This marked the beginning of a career with the provincial government in which he was to become the second provincial forester and a deputy minister. *Two Billion Trees and Counting* is an account of the contribution that this modest forester made to Ontario and its landscape.

Zavitz recognized that the only way to reforest was with species that could grow on soils with little to no organic content and withstand drought. The pines, especially Red Pine, were the ideal and natural choice. As a forester, he knew that what had to be initiated was a succession in which the pines would be the primary step. Once the crowns of these trees closed, subsequent forest management practices such as thinning could be undertaken, resulting in the development of a new forest rich in many species of plants and animals.

It seems strange, but as those pine plantations grew through their first two to three decades, the foresters who established them — and by implication Zavitz — were excoriated for creating sterile monocultures. Modern-day restoration ecologists would do well to learn from Zavitz's work: those "monocultures" are now providing major recreation and ecological value as well as contributing to society through timber and other products. A visit to any of the municipally owned-and-managed forests established during the early and mid 1900s provides tangible evidence of what reforestation can accomplish — the realization of Zavitz's vision.

Some years ago, *The Man Who Planted Trees*, a story about a French shepherd, Elzéard Bouffier, received much attention as a true account (it was later made into a film), but was later acknowledged by the author Jean Giono to be fictional. While the legend of Johnny Appleseed is ongoing in America, there are real-life promoters of tree planting who have been recognized. The most recent is Wangari Maathai, founder of the "Green Belt Movement" and recipient of the Nobel Peace Prize in 2004 for her efforts in sustainable development, democracy, and peace. Canada, and Ontario particularly, should be proud of the efforts of Edmund Zavitz, who more than a century ago set about the reforestation of wastelands, resulting in the forests that are now living monuments to his work. He was truly Ontario's "father of reforestation."

Kenneth A. Armson, President
The Forest History Society of Ontario

Acknowledgements

My book has been helped in numerous ways by environmentalists. In particular, two foresters, Ed Borczon and Dolf Wynia, played major roles in bringing the book to completion. They generously supplied photographs and carefully reviewed the entire text, pointing out what would have been otherwise embarrassing errors and shortcomings. Without Ed's patient encouragement, the whole project might have been abandoned; it was a great pleasure to meet someone who fully understood the reality of how Edmund Zavitz rescued Ontario through the planting of trees, and who has spent much of his life telling this compelling story. Dolf Wynia generously provided me with all the back issues of the remarkable magazine *Sylva* and numerous other historical manuscripts.

It was a delight to work in the inspiring St. Williams Forestry Interpretive Centre, which Zavitz played such a critical role in creating, and to wander beneath the towering pines he planted over a century ago. The intense dialogues we had there have shaped this book and increased my respect for Edmund Zavitz's achievements.

Many other environmentalists concerned for our province played important roles. My friend Albert Garofalo drove me to St. Williams in 2008 to commemorate the 100th anniversary of the St. Williams Reforestation Station. Here, I met the chief forester of Ontario, Ken Armson. His wise words and encouragement have been of enormous help in turning this biography into a reality. Carla Vonn Worden and Marianne Yake, of the Richmond Field Naturalists, both provided photographs and fostered my first public-speaking engagement on the life of Edmund Zavitz. Brenlee Robinson helped me with her inspired forestry thesis, while Caroline Mach of the Dufferin County Forests and Don Pearson of Conservation Ontario provided much encouragement. Eleanor Heagy of the Upper Thames Conservation Authority kindly provided photographs.

Two years ago, Ken Armson made the helpful suggestion that I write an article on the life of Edmund Zavitz as a prelude to a biography. The editor of *Buffalo Spree Magazine*, Elizabeth Licata, was supportive of the idea, and my first piece of writing about Edmund Zavitz was published.

In addition to the St. Williams Interpretive Centre, a number of archives and museums in Ontario have assisted with my research. I am grateful to the staff of the University of Toronto Archives for their help in locating critical writings by Zavitz and an impressive clippings file on members of the Faculty of Forestry. Much of book's core comes from the collections of the Archives of Ontario, including E.C. Drury's papers, the records of Zavitz's Forestry Protection Branch, and, of course, their remarkable photo collection.

Jude Scott of Fort Erie Museum Services guided me through their photo collection and files on the Zavitz family. The museum's records put me in touch with two of Zavitz's grandchildren, Kathleen Mackenzie and Peter Zavitz. They provided personal insights into the life and struggles of their grandfather, which cannot be found in the published records. I thank them for their co-operation. Among the many resources consulted were the writings of the late Clarence F. Coons. His work contributed much toward my understanding of the world of forestry in Ontario.

A special thank you must go to Dave Lemkay, the former general manager of the Canadian Forestry Association, for his encouragement and support, and to Jim Coats, the retired executive vice-president of the Ontario Forestry Association, for his support and assistance in acquiring images. Jan Brett and Darlene Wiltsie of the Archival and

— *Acknowledgements* —

Special Collections, University of Guelph Library, are thanked for their exceptional co-operation in locating an image and getting to it to us in record time. Both the County of Simcoe and Graeme Davis, county forester, are thanked for their co-operation in making images available for this book. The interest shown by Caroline Mach, country forest manager for the County of Dufferin, is reflected in the advance publicity printed in spring issue of *The Professional Forester*. Thank you, Caroline.

My own family played a crucial role in this project. My wife, Mary Lou, toured and photographed the astonishing forest resurrections of the Ottawa Region along with her brother Vincent Greason. They also helped track down the ravaged Rockland Plantation, which took two years of searching.

My friend Danny Beaton gave me an unexpected tour of the Hendrie Forest, created by Zavitz near Midhurst Ontario. The work of the Preservation of Agricultural Lands Society, through the efforts of the late Mel Swart and Robert Hoover, Jean Grandoni, and Gracia Janes, provided me with a greater understanding of the importance of Zavitz's role in fashioning some of their basic tools for protecting the landscape, notably tree protection bylaws.

Writers face the daunting challenge of inspiring readers. Here, I must give credit to the work of Dundurn's editor, Jane Gibson, who has guided me along the way. I thank her for dedication. I also thank Barry Penhale for his helpful suggestions and his belief in this project. To Matt Baker, my copy editor, I give grateful thanks for his very thorough attention to the overall readability of this work.

Introduction

My introduction to Edmund Zavitz came at the age of twelve, when I was fortunate enough to pick up a copy of *Renewing Nature's Wealth*, the history of the Department of Lands and Forests. Reading its inspiring story of Zavitz and his heroic battles to safeguard Ontario's natural ecosystems from the devastating effects of forest fires, spreading deserts, and rising floods reminded me of a phrase I had picked up from an admirer of the American pioneer environmentalist John Muir. Muir, who was much against the mad frenzy of greed that gripped California at the turn of the twentieth century, was referred to as the "only sane man in San Francisco." From my historical perspective, it seems that Zavitz, on the eve of the First World War, was a similar bastion of sanity in Ontario.

From my earliest years, my parents carefully instilled the wisdom of past conservationist achievements in me. My father Win was a high-school geography teacher, and, at the time of his marriage to my mother Mary, lived in Delhi, a town in Norfolk County. We took frequent family trips, and as we travelled, he would tell me about the time when much of Norfolk County was a desert wasteland and how the planting of beautiful forests dominated by White Pines that we could see around us had transformed the land. The ranch-style home he had built was surrounded by one such spectacular pine forest, the mature trees having been planted in the 1920s. During one trip around the Lake Erie shore, we stumbled upon the memorial plaque to Edmund Zavitz, the man who had made this transformation from desert wasteland to productive land possible.

A still-in-progress biography of Mel Swart, a key founder of the Preservation of Agricultural Lands Society (PALS) and with whom I work, led to *Two Billion Trees and Counting*. Swart told me that what had turned him into a conservationist was reading the "Report of the Select Committee on Conservation," published by the Ontario legislature in 1950, which I, of course, dutifully and subsequently read. The report contained detailed description of Zavitz's actions in establishing county forests across much of Southern Ontario. To Swart's great frustration, such forests were not part of the landscape of the Niagara region where he lived.

Lands Swart wished to have reforested were being gobbled up at an alarming rate by avaricious speculators abetted by their political allies. The minutes of Welland County Council describe the struggles Swart had to go through to achieve a forest-protection bylaw and ultimately the establishment of the Niagara Peninsula Conservation Authority in 1959. One of the reasons he championed these causes was an appeal issued in 1956 by Monroe Landon of the Norfolk Chamber of Commerce, asking that Swart give support to reforestation throughout Southern Ontario. I later learned that Monroe Landon was a long-time friend of Edmund Zavitz. The county minutes also noted that in the midst of Swart's battles in the Niagara area, he had journeyed to St. Williams, Ontario, in Norfolk County, to seek advice and meet with Zavitz and Landon.

That Edmund Zavitz, in his eighties at the time of the visit, had been such a force for conservation in the 1950s was a surprise for me. I had generally associated him with conservationist struggles that climaxed in 1934 when the Hepburn government forced him from his role as deputy minister of forests. He was then appointed director of the reforestation branch within the department and his powers restricted to Southern Ontario only. My admiration for Zavitz deepened as I learned more from Swart.

— Introduction —

I later came to understand that the reason why reforestation measures and funds for conservation authorities were slashed so horrifically in the 1990s during the Bob Rae government was that there was a very limited public understanding of why these programs were created. Very few people in Ontario understood how much of the province had been degraded to desert-like wastelands, largely caused by extensive deforestation. Likewise, there was little understanding that loss of forest cover was related to the massive flooding (this sparked the creation of conservation authorities). Even less known was the danger of man-made forest fires burning northern soils down to bare rock, creating alvar-like conditions and leaving insufficient soil to support the growth of trees and most bushes.

Taking part in numerous public consultations on behalf of PALS also increased my determination to write about Zavitz and his work. During these sessions, I often discovered that some people used strange reasons to challenge the formation of a greenbelt. One, that its boundaries were determined by "political science," totally ignores the greenbelt's obvious intention: to protect the watersheds that flow from the headwaters of streams that originate in the Niagara Escarpment and Oak Ridges Moraine. Another, an unsubstantiated concern, was that the Greenbelt would become a zone of estates for the rich. The reality is that the protection of watersheds from deforestation and development sprawl will permit people to walk on trails and enjoy fish from healthy streams.

My coming to understand the importance of the tree bylaws Zavitz created, which were enforced by professional foresters, was another motivation for delving into research and writing this book. Interestingly, the employment of professional foresters has enabled our society to protect threatened urban forests. However, those forests found on the frontiers separating cities from farmlands are at the mercy of zoning battles that involve developers, municipalities, and conservationist groups.

By the time I went to St. Williams in July 2008, to attend the 100th anniversary of the St. Williams Reforestation Station, I had already resolved to write a biography of Edmund Zavitz. The beauty of the forest there, resembling an old-growth forest similar to the pines of Oka, increased my determination even more: we need to appreciate Edmund Zavitz and his dedicated work in reforestation. Perhaps through understanding his methods, Ontario can once more become a leader in the ecologically responsible care of lands and forests.

Edmund Zavitz was a remarkable person. Ontario, and all of its citizens, owe him greater recognition. I feel honoured to have had the opportunity to record his many accomplishments, and it would seem most appropriate that this long-overdue biography make its appearance during the UN's International Year of Forests.

— ONE —

The Man Who Did Plant Trees

THE SUMMER OF 2010 brought a barrage of global environmental disasters, bringing to mind Ontario of the early twentieth century, when the province had provided the stage for ongoing environmental disasters and loss of life. It was this type of scenario that propelled Edmund John Zavitz into public view. His mission: to rescue Ontario from further devastation. In 2010, massive forest fires in Russia engulfed Moscow in clouds of smoke that kept millions indoors, in fear of their lives, and threatened the security of nuclear power plants. Pakistan was gripped with a wave of floods that made millions homeless, unleashed landslides, devastated crops, and brought deadly epidemics.

Faced with this onslaught of foreign disasters, not one Canadian media pundit remarked on how these scenes of horror, then the staple of news coverage, were once commonplace in Ontario. From the late1880s to the 1920s, forest fires destroyed Northern Ontario towns so frequently that losing one's home to a fiery inferno was seen as the price of living in the booming frontier. Floods that once engulfed communities such as London on the Thames, and Port Hope on the Ganaraska River, were as frequent as those of Karachi and Lahore, victims of the Indus River.

Much like the bleakest reports out of Africa today, describing the relentless march of the Sahara, descriptions of Ontario included the threats of two kinds of spreading deserts. To the north, rock deserts, much like the later moonscape of Sudbury caused by sulphur emissions, were expanding, the result of repeated forest fires burning away the thin soils of the Canadian Shield. To the south, deserts created by blow sand, caused by erosion, engulfed apple orchards, factories, and roads.

During his childhood, Edmund Zavitz's mother, Dorothy, and his grandfather, Edmund Prout, told him stories about the devastation along the Oak Ridges Moraine. Some years later, in 1909, Zavitz photographed this scene of extensive erosion at the headwaters of the Ganaraska River near their farm, and lamented, "A river started here."

Those who were gripped by the mayhem of the news broadcasts of 2010 might have said, "There, but for the impact of Edmund Zavitz, goes Ontario." His tripling of forest cover in Southern Ontario ended the threat of the marching deserts, which were termed "blow-sand conditions," on lakeshores and moraines. His efforts were assisted by the introduction of new fire regulations, which he crafted, across all Ontario Crown lands, thus ending the blazing infernos that consumed soil cover in the boreal forest.

Zavitz's rescue of Ontario from ecological degradation was a very long process, beginning with his employment as a graduate student in 1904, and ending with his death in retirement in 1968. His forestry reforms were based on regulations developed in Imperial India during the mid-nineteenth century through the creation of the Indian Forest Service. The regulations, which prevented clear cuts deemed too large to encourage natural regeneration and restricted livestock from grazing in forests, had been met with fierce resistance from vested logging interests seeking short-term profits. This bitter conflict became a drawn-out battle that was not resolved in Ontario until the 1940s.

This photograph, labelled by E.J. Zavitz as "Ganaraska Wasteland Photo #23," reveals some of the humour that helped him mobilize his reforestation efforts.

In their comprehensive history of the Ontario Department of Lands and Forests, historians Paul Pross and Richard Lambert use the term "legendary figure" to describe E.J. Zavitz.[1] As these historians acknowledge, however, this term represents the way Zavitz was viewed in the last three decades of his life, the 1940–60s, by those who were employed by the Department of Lands and Forests. They viewed Zavitz as a powerful role model, one they emulated to excel in their work. Generally, with the exception of Norfolk County, which he also rescued from the threat of drifting sand, his work had limited public exposure.

When Pross and Lambert described Zavitz as a living legend, his policies had become "holy writ" for senior civil servants and parliamentarians of all parties in Ontario. This reality was movingly acknowledged in the then-premier John Robarts's introduction to their book. Here, he praised not only Zavitz but the entire staff of the department, who were inspired by him. Zavitz appreciated his staff, recognizing them as well-trained and "authorized to take action" to combat serious threats like floods and forest fires.[2] When Robarts wrote his introduction, reforestation efforts were at their peak. Twenty-million trees were being planted annually, and ambitious tasks such as reforesting much of the Ottawa Greenbelt, which had been through astonishingly bold expropriations by the federal government, were underway.[3]

By the 1980s, under Robart's successor William Davis, reforestation projects were on the decline. Provincial funding allowing conservation authorities (one of the most remarkable legacies of Edmund Zavitz) to acquire properties was being cut, significantly reducing their ability to acquire environmentally sensitive lands important for watershed protection, such as the Oak Ridges Moraine.[4]

While Zavitz's watershed protection goals of reforestation in Eastern Ontario had been largely achieved by the 1980s, with an average rate of 38 percent forest cover, this was not the case in southwestern Ontario. Here, streams still suffer from flash conditions, becoming floods in the spring and bone dry the rest of the year, except when a big storm occurs. These cycles of floods and drought conditions lead to fish-habitat loss. Today, in Essex County alone, barely 5 percent of the total land has forest cover.[5]

Zavitz and the professional foresters who worked for him and whom he hired directly, such as his long-time assistant, A.H. Richardson, were well-schooled in the complexities of Ontario's natural and human history. Both Zavitz and Richardson would encourage the conservation authorities they created to establish historical interpretation programs, like Black Creek Pioneer village. They wanted people to understand the impact pioneer practices had on the environment — like cutting down trees to make soap — and what could be done to rectify the damage. Using his extensive knowledge, Richardson ensured that new conservation authorities emphasized the teaching of watershed history to the general public.[6]

Zavitz's reforestation work so improved the Ganaraska that today it is one of the best-forested watersheds along Lake Ontario. The beauty of this rural scene, showing the restored Ganaraska Forest, is one of the reasons that citizens do battle today to protect the landscape of the Ontario Greenbelt.

Zavitz's success in Ontario was not a one-person miracle; his extended family played an enormous role. His father-in-law, John Dryden, gave him a political boost, and his wife, Jessie, provided exceptional support. Zavitz and some members of his team of professional foresters were also likely the only people to be close friends with the two most-bitter political adversaries of the 1920s — Ernest C. Drury (coalition government of United Farmers of Ontario and Independent Labour Party 1919–23) and Howard Ferguson (Conservative, 1923–30), who both served as premier of Ontario.

Zavitz's personal style enabled him to launch popular movements, one of which led to the Conservation Authorities Act of 1946. This act, until the mid-1980s, provided 50 percent of the funds that the Ontario Conservation Authorities required for land purchases. Between 1937 and 1946, Zavitz became a great grey figure of eminence, operating in the background.[7]

In the late 1990s, the government of Mike Harris unleashed the Common Sense Revolution. As a result, provincial funding to conservation authorities was slashed, impeding Edmund Zavitz's legacy of restoration work and reforestation projects. This funding reduction forced the sale of some previously protected forests and eliminated provincial reforestation stations. In encouraging this, Harris was radically accelerating trends endorsed over the past fifteen years by the governments of all three political parties — even the New Democratic Party participated. The party's premier, Bob Rae, took to insulting public servants involved in forestry by suggesting they were so far behind the times in their practices that they were still using quill pens. Presumably, he was unaware of their work in building on Zavitz's reforestation mission. Typical of this government's myopia, they extended the Tile Drainage Act of 1878 to allow the draining of swamp wetlands to Indian Reserves, but did not extend Zavitz's Agreement Forest Program that had formerly repaired the damage this draining caused.

Although crippled and made more expensive by the demise of Zavitz's greening efforts, it appears that reforestation measures may, once again, be on the rise. One such program is the connecting of J.H. White Forest (adjacent to Turkey Point Provincial Park) and the Edmund Zavitz Forest. Another program involved purchasing land for new conservation areas on the Oak Ridges Moraine through the creation of a provincially endowed Oak Ridges Moraine Foundation. The new Robert Hunter Provincial Park, near Markham, named in honour of the founder of Greenpeace, is yet another. Here, the intent is to implement some of the reforestation envisaged in A. H. Richardson's watershed plan for the Rouge River devised some fifty years ago.[8]

Climate change makes Zavitz's ideas even more pressing today than they were in his lifetime. When the British imperial forestry practices he championed first began in India, foresters everywhere came to understand how deforestation could harm rainfall patterns and damage watersheds. Today, regional climate-change issues have assumed even more frightening global dimensions, most ominously affecting the chemical composition of the oceans. Although the exact role of forests in carbon sequestration — the ability to store carbon from the atmosphere — is subject to great scientific debate, the need for more forest cover to protect watersheds from even the most optimistic predictions of climate change is greater than ever.[9]

Similar to the American influence of forester Gifford Pinchot, the creator of the U.S. Forest Service, Zavitz's career represented a major turning point in Ontario. Pinchot's influence on the long-term health of American forests and the watersheds they protect has been vividly termed by his biographer, Char Miller, as "the making of Modern Environmentalism."[10] In a similar vein, Zavitz nurtured the birth of the environmental movement in Canada. The movement in Ontario, however, was a considerable challenge; Zavitz, though exceptionally competent and dedicated, lacked the wealthy patrician support enjoyed by Pinchot through his close friend, President Theodore Roosevelt.

Interestingly, despite Zavitz's remarkable achievements in the face of an uphill battle, his name remains relatively unknown. On the other hand, Americans have been more celebratory of Pinchot's legacy. His early reforestation projects at his family home of Grey Towers, Pennsylvania, and at the former Biltmore Estate in North Carolina, are still vigorously protected as national shrines. The Canadian folly of not valuing its environmental exemplars is vividly demonstrated by the blatant disregard for the country's, indeed the continent's, first successful reforestation in Oka, Quebec. In the summer of 1990, the site became the scene of a bitter battle between the Mohawks of this community and the municipal government. Plans to cut up a restored forest of towering White Pines for a combined residential development and golf course sparked a crisis situation. In contrast, Pinchot's projects are carefully studied for information on the long-term management of White Pine, since old-growth stands are incredibly rare in the United States.[11]

During the nineteenth century, reforestation was the subject of intense debates, similar to those in our own day about global warming. The impending demise of the valuable White Pine forests across North America was shrugged off by most logging interests, with the view that such a fate was inevitable. These assumptions were shaken by the actions of the Sulpician Order priest Father Joseph Daniel Lefebvre. He, with the help of his Mohawk parish, planted some 65,000 spruce and White Pine seedlings in 1888. They obtained the seedlings from the forest, unlike later techniques by employed nurseries, which use seed extracted from cones. In the process, they rescued the community of Oka from spreading deserts that earlier had buried the town in an avalanche of sand.[12]

The importance of this process was well-captured in the remarks by J.R.M. "Mack" Williams, an award-winning Ontario forester, in a 2006 Carolinian Forestry conference on the future of the landscape of Southern Ontario. He observed:

> The late Henry Koch of the University of Guelph Arboretum often noted that the spread of civilization around the world has too often been followed by the spread of deserts. He noted that a century ago we were well along that road in Ontario, then perhaps for the first time in history, humans turned back at the brink and the remarkable effort at reforestation, forest management and other conservation efforts followed.[13]

The person responsible for this first turning-back from the brink of destructive folly was Edmund Zavitz — the man who did plant trees.

— Two —

Early Influences

Edmund John Zavitz was born on July 9, 1875, in the village of Ridgeway in Bertie Township, a municipality annexed into the Town of Fort Erie in 1970. His birthplace, at 477 Ridge Street, is a distinguished red-brick Italianate home that today is still used as a residential dwelling and is also designated as a historic property by the Town of Fort Erie. Its grounds are still graced with the White and Red Pine trees he planted there — a feature rare in a community that had its coniferous giants stripped away long ago by avaricious loggers. Young Zavitz spent his formative years here before heading off to the more distant Woodstock Collegiate in Woodstock, Ontario, McMaster University (then in Toronto), Yale, and the University of Michigan, graduating from the latter with a Master's degree in forestry.

Although born into an "age of fire" where a "burn, baby, burn" philosophy expressed the prevailing attitude toward forests believed to stand in the way of progress, young Edmund was fortunate to know some people in Bertie Township who thought differently. They influenced him when he was a boy and helped him develop an interest in studying and exploring the natural world. One influential individual was the unusual conservationist Peter Shisler, a most colourful intellectual found behind the plough — he was a farmer and inveterate writer of letters to newspapers. In May 1872, he had argued for farmers being required to have a portion of their lands reserved in forest cover.

From Shisler's perspective, many were simply pretending to be farmers to "only live on the timber and then leave."[1] His oft-expressed concern became more public when the *Welland Tribune*, the newspaper closest to Bertie Township, published an editorial on June 21, 1961, and used Shisler's words to explain the need for more forest cover in Welland County. Some years later, the paper lamented that Shisler's vision was "eighty years ahead of his time," and bemoaned the fact there was now "a bylaw to regulate tree cutting but only after the lands in the county were denuded of all but about 5 percent of bushland" — far below the amount seen as necessary by conservation authorities.[2]

Nurtured by Zavitz's ideals, Mel Swart became a "conservation activist." During the 1950s, while serving on Welland County Council, he promoted Zavitz's conservation ideas in the Niagara area through the creation of the Niagara Peninsula Conservation Authority and the implementing of tree-cutting bylaws. Along with his concern for the reforestation of Southern Ontario, Swart also played a critical role in the establishment of the Niagara Escarpment Plan, the first link in Ontario's Greenbelt. This photo was taken in 2003 at an unveiling of a plaque to celebrate the recognition of the Niagara Escarpment Biosphere Reserve.

The *Tribune's* editorial illustrates just how revolutionary Zavitz's ideas for environmental protection were when he first began his work in the early twentieth century. Somehow, against all odds, he managed to transform practices that were considered odd or unnecessary into the accepted way of thinking by the time he retired. Actually, both the Welland County Tree-Cutting Bylaw and the Niagara Peninsula Conservation Authority were creations of the then-warden of Welland County, Mel Swart.[3] When preparing the bylaw, which was cited approvingly by the *Welland Tribune,* Swart had turned to Zavitz for advice. By this time, Zavitz was in his eighties, living on the family farm then being run by his son Ross, in Forestville near St. Williams. Zavitz was still the person to approach when it came to guidance regarding trees.

In Ridgeway, Edmund's father, Joseph, held the respected position of postmaster, the post office being located right in the Zavitz General Store. Young Edmund, as might be expected, spent much time there helping his father. In those days, the general store was frequently the community hub, where people not only shopped and retrieved their mail, but also stayed to exchange ideas and gossip. Over the years, Edmund would have been exposed to a number of people, including those whose ideas lay outside the accepted practices. With his keen interest in the natural world, he would have been particularly attuned to those "mavericks" who believed in protecting forests — an idea that was far different from the norm of the community. This divergent way of thinking about trees would have been reinforced by his mother, Dorothy, who had grown up on a farm that was partly on the Oak Ridges Moraine. Here, as conservation authority reports would document, erosion had caused so many dust storms that women were careful to delay spring cleaning until vegetation sprang up.

In his "Recollections, 1875–1964," Edmund Zavitz wrote, "my grandfather was concerned for the future owing to the destruction of the woodlands." Dorothy explained to her son that grandfather Edmund Prout's farm was on the borders of the Ganaraska watershed, and had "portions of sandy soil, which upon being tilled deteriorated."[4] His grandfather, Edmund Prout, had been born in Devonshire, and emigrated from England with his family in 1852. They settled on a farm on the Oak Ridges Moraine in Clarke Township when Dorothy was only two years of age.

Prout would often take his young grandson on walks through the woods and teach him how to observe things carefully, things that most other people would not notice.[5] In an interview later in life, Zavitz recalled these walks taken when he was "but a blue-eyed Victorian tyke in faded blue shorts." He remembered how grandfather Prout "took him on tours through the local orchards, telling [him] about the miracles of trees."[6] Edmund's mother, being steeped in the conservation ethics of her father, maintained her zeal for restoration of the forests and would be a strong supporter of her son until she died in 1941 at the age of ninety-six. She thoroughly understood and encouraged his conservationist ideals.

After his grandfather died in 1897, Edmund kept in touch with other maternal relatives who continued to live on the farm on the headwaters of the Ganaraska River. These included his uncle, John Squair,[7] a French professor at the University of Toronto; Laura Prout, one of his mother's sisters and the wife of John Squair; and Francis Squair, who inherited grandfather Prout's farm.

Uncle John, drawing on his personal experience concerning the need to reforest the Oak Ridges Moraine, would, like Dorothy, also be a constant source of support for Zavitz. In 1926, John Squair's book, *The History of Darlington and Clarke Townships*, was published by the University of Toronto Press. Much appreciated for its details on local ecology, it continues to be cited by the conservation authorities along the Oak Ridges Moraine as a source for understanding its ecological history.

Squair agreed with pioneer author Susanna Moodie (1803–85), who saw the burning of forests by farmers to clear land as a diabolical ritual. She conjured up a vision of the devil merrily fiddling above burning heaps of logs from trees cut down during a drunken logging bee. Squair wrote:

— *Early Influences* —

In the beginning there was sometimes a feeling that a tree was rather an enemy than a friend; a thing to be rooted out and burnt up. The burning of the log-heaps in the evening after the day's logging was done was an occasion for rejoicing and for passing around the whisky jug. It might well have been a season of regrets. The natural results followed. The thinning of the forest began, and the human agencies were not infrequently, aided by the forces of nature. Tradition and the newspaper have kept alive the memory of forests destroyed by high winds like the tornado of July 12, 1850, when in Cartwright and Darlington, houses, barns, and woods were destroyed throughout a strip of territory of

Name of photographer not known. Courtesy of Ed Borczon.

This posed photo of men involved in a "chopping bee" was taken in 1899, near the Belleville home of the Canadian pioneer author Susanna Moodie. Both she and Zavitz's uncle, John Squair, who seemed to be familiar with her writings, regarded logging bees as diabolical rituals that celebrated the death of forests and frequently culminated in violence and drunkenness.

considerable length. Then into the "slash" caused by the hurricane, would drop, in the following summer perhaps, a spark of fire which would spread into flame, and carry devastation for long distances, as was the case in the Ridge [Oak Ridges Moraine] about June 1, 1855, when the barn of a Mr. Campbell was burned, his mill being saved by the most energetic exertions.[8]

Uncle John explained to Edmund that the land on the farm, which was now a wasteland desert, had once been an example of "a remarkable rich variety of the gifts of Mother Earth." The woods had provided the material for great craftsmen, whom he recalled flourishing during his childhood.[9] Squair regretted his role in turning the Moraine's forest into a desert. He recalled how the land he had cleared was "swampy, some of it gravelly and covered with big stones." He went on to describe the destruction of the land:

> Then most of it had been stripped of its valuable lumber — cedar, pine, beech and maple — but there remained a good deal of the rougher cedar, hemlock, elm, basswood, which it fell to my lot to cut down and work into fine wood and rails, and team away. A large amount was burned in log heaps after the wood and rails had been taken off.

While he briefly experienced the sensation of "the joy of tilling the virgin soil," Squair found soon that this happiness was an illusion. He quickly discovered "the land was poor and rarely did we get good crops."[10]

Notwithstanding his mistakes, Squair took comfort in knowing that his family had become among the most ecologically conscious farmers in Darlington Township. He understood that while "the majority did not worry about the forests," there were a few owners of property who "began to be careful of wooded areas." Included in these were "Edward Rutledge, and his neighbour in the community of Salem, Zavitz's grandfather Edmund Prout." From Squair's perspective, his father-in-law Prout, whose farm was located at Lot 7, Concession 3, Darlington Township, was "the pioneer of sylviculture in Darlington." Consequently, he found it appropriate that here "one can see on his old farm the first pine plantation in the Township, and in addition, the Province has his grandson, E.J. Zavitz as its ... forester."[11]

Squair also noted that by the time Zavitz was attending elementary school, the attitude of "war" on nature that had characterized the frequently drunken logging bees of the past had begun to change. In 1885, when Edmund Zavitz was ten, he was taking part in provincially supervised Arbor Day celebrations, which, he noted, had been established "in response to the growing feeling in favour of planting trees."[12] John Squair lived until 1928 and was able to take considerable pride in his nephew's accomplishments. By the time of his uncle's death, Zavitz had begun the Moraine's reforestation on a massive scale.

In addition to the extensive influence from his maternal relations, Zavitz also benefited greatly from Bertie Township School No. 11, the Continuation School in Ridgeway, where the teacher and amateur naturalist Alva Kilman was principal.[13] An exceptionally dedicated teacher, he not only founded the township's first public library but also transformed School No. 11 into the first continuation school to offer classes up to grade ten in Bertie Township. His students did so well, graduating with grade ten leaving certificates, that the school received provincial bonuses as a result. Unlike the short-lived pattern of previous teachers in Ridgeway, Kilman taught there for eighteen years.[14]

Edmund Zavitz saw Kilman as "a remarkable naturalist of the old school." Kilman, Zavitz also noted, "was also an ornithologist and taxidermist, having collected and mounted a very fine collection of birds. Fortunately for Kilman, the Niagara Peninsula was one of the points in line with the migratory flights."[15] In later years, Kilman's collection was acquired by the Ontario Agricultural College (OAC).

Kilman further nurtured young Zavitz's love for the natural world through expeditions into the woods, usually in the company of his son Leroy, who would become Edmund's life-long friend. As a result of these walks in the woods of Niagara, Zavitz, now well on his way to becoming an amateur naturalist, continued to collect moths, butterflies, and beetles into adulthood. When he decided to go to Yale to study forestry, Zavitz turned his collection of butterflies and moths over to OAC and kept only his beetle collection connected with woods and forestry subjects. This collection of beetles changed hands a couple times and eventually found its way to the Royal Ontario Museum, where it is now part of a display case. Edmund graduated in forestry from the University of Michigan. His friend Leroy also received his degree in law at the same university.[16]

With his rambles in the woods searching for migrating hawks, moths, beetles and butterflies, and excursions with his grandfather, Edmund Zavitz's early childhood was a happy one. His home, his school, and his father's post office were all within a few blocks of each other on Ridge Street. Ridgeway, then a community of about six hundred people, was a pretty village. With much of its business activity connected to the neighbouring resort town of Crystal Beach on Lake Erie, the town was at its peak of popularity as a get-away destination for Buffalo residents.

In addition to the positive influences from his maternal family and the Kilmans, Zavitz benefited as the son of the village postmaster, Joseph Zavitz. Joseph was the descendant of a distinguished Quaker family, and they had a record of high regard for education going back to the family's European origins in early eighteenth-century Strasbourg, the capital of Alsace. This had been the centre of Protestant learning and culture in Europe before its conquest by Louis XIV of France. The trauma of the French conquest of Protestant- and German-speaking Alsace caused George Zavitz to leave Strasbourg for Philadelphia in 1727, whereupon he soon became a Quaker. After the decline in the political power of Quakers in Pennsylvania following the American Revolution, three of his descendants — the brothers, Henry, Jacob, and Christian — settled in the southern Niagara region. Joseph was the grandson of Jacob Zavitz, who had settled in Bertie Township in 1784. One of Jacob Zavitz's first steps upon arriving in Bertie Township was to found a Quaker school.[17]

While Jacob Zavitz's family stayed in Niagara, the sons of other brothers moved to Western Ontario, where they established Quaker settlements in London, Sparta, and eventually Coldwater near Georgian Bay. Among these Quakers who moved west was another prophetic figure, Charles A. Zavitz, a man whose flowing beard and strong sense of justice seemed directly out of the pages of the Old Testament. He had grown up in Coldwater, then later moved to London, and demonstrated a combination of high intellect and principle similar to his third cousin, Edmund Zavitz. Charles was the founder of the Canadian Peace and Arbitration Society, Canada's first peace group, and a professor of agriculture at the Ontario Agricultural College (OAC). Here, he refused to allow an armed militia to maintain a presence during the First World War, because of his Christian convictions.

Charles was one of the leading agricultural scientists in Canada at the time and had introduced soybeans as a crop for Ontario agriculture. He also worked closely with the Fruit Growers of Ontario. This particular branch of the Zavitz family became pioneer advocates of reforestation after the deforestation of Southern Ontario had brought greater cold and winds, damaging fruit crops. Charles Zavitz incorporated the cooperative, innovative agricultural experiments conducted by members of the Fruit Growers into his work. He also established the Experimental Union, an alumni association of former OAC students. Its members tested Charles Zavitz's ideas, such as the introduction of soybeans. The Union became directly responsible for the hiring of Edmund Zavitz to run the province's first tree nursery for reforestation by farmers in 1904.[18]

Charles Zavitz was a close friend of William Brown, who had emigrated from Scotland in 1871 and acquired a farm near Orillia. Initially, he worked as provincial land surveyor. William was the son of James Brown, a leading British forester who had written an influential textbook on forest management in 1863 and was one of the early foresters to recognize the negative impact that deforestation had on the environment. He earned a gold medal in Scotland in recognition of his mastery of this controversial topic. William, like his father, had chosen a career in forestry and became a manager for great estates in Scotland, where he undertook extensive reforestation. While there, he received a gold medal from the Highland and Agricultural Society of Scotland and the Scottish Arboricultural Society for his exemplary work on the topics of forestry and climate. Once established in Canada, he became a professor of agriculture at the Ontario Agricultural College in Guelph in 1874. He is the person who established the university's Arboretum and transformed a gravel pit into the forest now called Brown's Woods, which is still on the Guelph University campus. William Brown also helped to encourage Ontario's first provincial forester, Judson Clark, an OAC student from Prince Edward Island, to take up a career in forestry.[19]

The happy world of Edmund Zavitz's childhood, shaped by so many diverse, positive, and loving influences, was disrupted by his father's death in September 25, 1887, at the age of seventy. Edmund was twelve years old. For two years, he stayed in school, going as far as was possible in the Bertie Township Continuation School. In "Recollections, 1875–1964," Zavitz indicated that he, at the age of fourteen, had no plans for further formal education, since he "was not interested in schooling." He first went to work, in 1889, on his uncle's farm on Black Creek, a tributary of the Niagara River, about ten miles from Ridgeway. This, however, was a short stint. For the next six years, he worked for the local and very successful entrepreneur Eber Cutler. His influence in Ridgeway can be seen in the name Cutler Street.

Edmund Zavitz describes Cutler as "the mogul who owned most of the town — the village general store, the sawmill and planing mill and the flour and feed mill, and who did contracting to build houses in the region." At one stage of his career, Cutler also owned the railway that took tourists from Ridgeway to Crystal Beach.

While in Cutler's employment, Zavitz "worked in the village bakery, [spent] one year as a plumber's helper, painted, [was a worker in a] flour and feed mill, and [a] clerk in the general store." The clerking position was acquired shortly after taking a six-month business-banking course in Buffalo. Zavitz was becoming a sort of "jack of all trades," gaining experience that would prove most helpful in later years, especially in designing particular equipment for photography.[20]

For a period of time, Zavitz worked in varied capacities for local mogul Eber Cutler. A young Edmund Zavitz is shown here with a group seeking to attract shoppers to Cutler's Dry Goods store in Ridgeway. Shortly after this photo was taken, his parents persuaded him to complete his high school education.

A new stage in Edmund Zavitz's life began, following his widowed mother's remarriage to I.L. Pound in 1892. In "Recollections, 1875–1964," Edmund Zavitz wrote, "[Pound] became a very good father to me and I revere his memory." Three years after their marriage, the Pounds had a serious talk with their son about his future (he was now twenty years old). They convinced him to attend St. Catharines Collegiate and complete grade eleven. However, headmaster W.J. Robertson's traditionalist views on education, which opposed nature study, did not inspire Zavitz to continue there. He completed grades 12 and 13 at Woodstock Collegiate, matriculating in the spring of 1898, when he was twenty-three. He then entered McMaster University (Toronto), and received his BA in 1903 at the age of twenty-eight.[21]

At McMaster, Zavitz came under the formidable influence of one of the Baptist school's critical founders and supporters: the minister of agriculture, John Dryden. He also met Dryden's daughter Jessie, who would later become Zavitz's wife. She and her sister Elizabeth were among the first women in Canada to earn college degrees. As was common among highly educated women in that era, Jessie poured a lot of energy and talent into church missionary projects. She taught mathematics for four years at the Baptist Moulton College and served as the alumni representative on the McMaster University Senate. Suffering from partial deafness (a problem shared by Zavitz), she became the founder of the Canadian Federation of Lip Reading Clubs.

John Dryden served as minister of agriculture in three Liberal governments for a total of fifteen years. A farmer and a believer in scientific agricultural practices, Dryden won a number of awards for his breeding of Shorthorn cattle. His farm-management methods were modelled on the recommendations of the Ontario Agricultural College's remarkable faculty, which included William Brown and Charles Zavitz. His scenic Maple Shade Farm was located close to Whitby in Ontario County, now part of Durham Region. Dryden planted trees and kept livestock in fenced pastures so they would not graze on and destroy natural forests. He would have had an ongoing relationship with Charles Zavitz for twenty years by the time he met Edmund Zavitz.[22]

From his childhood in Ridgeway to his later teaching years in Guelph, Edmund loved to take part in a variety of sports. At McMaster he was captain of the soccer team when the university won the Caledonia Cup, the championship for the western half of Ontario. He also played on championship hockey teams and took part in rugby games while teaching at the Ontario Agricultural College in Guelph. One of his teammates was the college half-back and star-player John Bracken. During their association, Bracken did pick up some of Zavitz's conservationist messages. After serving as professor of agriculture, Bracken became premier of Manitoba for twenty-two years. Under his stewardship, the conservation regulations of the federally-based Canadian Forest Service would be transferred to provincial governance in until 1934.[23]

A critical turning point in Zavitz's life came in 1903, his last year at McMaster. A biology professor, Richard Smith, gave him a pamphlet issued by the newly formed U.S. Forest Service, then under the inspired direction of its conservationist first administrator, Gifford Pinchot. In "Recollections, 1875–1964," Zavitz recalls how at this moment his life's direction was set: "Up to that time I had no idea of the future unless it was in the teaching field." He immediately inquired about North American forestry schools, selecting the one founded and heavily endowed by Pinchot, at Yale University.

About the same time that he was investigating various forestry schools, Edmund Zavitz attended a banquet in Toronto, sponsored by the Canadian Forestry Association, which was encouraging the introduction of a conservationist approach to forest management across the nation. Here, he met Thomas Southworth, who was the secretary of the Ontario government's Bureau of Forestry at the time, and later the director. He also met Judson Clark, who had recently moved to Toronto and was soon to be appointed first chief forester of Ontario. The two men would become good friends and visit forests across Ontario together, documenting their findings — often through photography.

Judson Clark's 1904 annual report to the Bureau of Forests gives a good account of the places that he and Zavitz explored together. One was Norfolk County. There, they took many photographs to document both exceptionally well-managed forest properties and the horrific consequences of sand-blown waste from deforestation. One such property was Backus Woods (this remarkable forest was recently acquired by the Nature Conservancy of Canada), one of the most biologically diverse remaining Carolinian forests. The photographs also showed the pattern of degradation of the remaining forest fragments — the result of livestock-grazing invasions. Zavitz later recalled, "We

Dr. Judson Clark, the first chief forester of Ontario, was photographed by Edmund Zavitz in 1903 while the two men were engaged in documenting aspects of Ontario's forests. The setting is Lincoln County, likely in the old growth forests of the Niagara Escarpment, which today are protected by the Balls Falls Conservation Area.

had a few interesting trips during his stay and he introduced me to photography as an aid to forestry work."[24] Zavitz's jack-of-all-trades talents helped him greatly. He needed technical agility to carry around the cart of equipment and carefully pose shots while sinking knee-deep in sand. He made his own lantern slides and glass negatives, crafted his own enlarger, and hooked up a special system for copying photographs. Zavitz would use this specialized equipment until 1967, a year before his death.[25]

Clark's and Zavitz's explorations put them into contact with farmers, whom they would attempt to persuade to plant trees. Regarding such early jaunts, Zavitz later told reporters, "I had travelled with horse and buggy, going from one farm to the next and demonstrating in person, just how trees should be planted to provide windbreaks, prevent erosion and even provide a paying Christmas tree crop. It required a great deal of talking, but then I always could talk a great deal ... Some of the farmers were reluctant to plant trees, but they quickly learned it was a paying proposition."[26]

During another photographic journey with Judson Clark, Edmund Zavitz discovered the sand deserts of Norfolk County for the first time. This is one of the dramatic photos he took of the great White Pine stumps found there. He used these images to mobilize public opinion for reforestation.

Apart from meeting the Drydens and being a sports star, Edmund Zavitz did not find McMaster University very stimulating. He struggled with a problem, chronic since he graduated from Continuation School in Ridgeway under Kilman's guidance, of a lack of interest in subjects that did not concern the natural world. In such circumstances, he must have resorted to last-minute cramming to pass most courses. While Edmund Zavitz did not develop many close bonds at McMaster, his characteristic politeness served him well in future years, and he managed to get alumni friends on side once his difficult mission of bringing a system of forest protection into Ontario was actually launched.

An image of Edmund Zavitz emerges in a charming tribute given by one of his former student colleagues at a banquet held in Barrie on January 8, 1948. The occasion was to promote a historical publication on Simcoe County, written by the former premier of Ontario, E.C. Drury. In his speech, A. Cranston, the prominent then-retired publisher of the *Barrie Examiner*, recalled how, at McMaster, Zavitz was known as "Fuzzy," a student whose "sole interest was in trees."[27] The speech recorded how his fellow students saw him at McMaster: "We considered him a bit queer in those days. He didn't really want to go to lectures or study for exams. He was always looking at trees."[28] While Cranston revealed that Zavitz appeared odd at McMaster, after graduation his fellow students began to appreciate the importance of his zeal. He concluded his speech by observing, "as a result of Fuzzy Zavitz's work, the province of Ontario is again being covered with forests, and we brag about having been to school with him."[29]

Cranston's remarks were illustrative of the patience Zavitz had exercised in carrying out the transformation of Ontario, shifting from regulating fire-bombing assaults on forests to establishing conservation practices. Eventually, his skeptics and critics were converted to conservationist thinking. While joking about "Fuzzy's" passion at McMaster for protecting nature, Cranston acknowledged that he too had been a skeptic, but later became dedicated to the cause. During the 1920s and 1930s, Cranston became the publisher of the most influential daily newspaper in Simcoe County, and, from this platform, supported the use of the Agreement Forest Program that Zavitz and E.C. Drury would develop to reforest thousands of acres of the formerly sand-blown wastelands in Simcoe County.

In addition to good manners, Zavitz had another effective technique to get converts for the cause of forest protection — a keen sense of humour. It was said that his "eyes twinkle[d] with humour." While it was not expressed in a cruel or offensive fashion, his humour was effective. He would turn his role as chair of banquets for the Department of Lands and Forests into one of a stand-up comedian. After being praised, for instance, for establishing the forest-fire protection system of Ontario, it was reported that he put the audience "into helpless laughter," when declaring, "You fellows think you are pretty smart with all your modern fire equipment and things. Why in the old days, we had bigger and better fires than you ever saw, and we didn't have all the fancy gadgets you have now."[30]

As Edmund Zavitz packed his bags to attend Yale in the summer of 1903, he was well on his way to a the role of rescuing Ontario from the threats of excessive forest fires, farmlands turning into deserts, and cities drowning in floods. Unknown to him, forest conservationists around the province, spearheaded by the Experimental Union and organized by his cousin Charles Zavitz, were carefully lobbying his future father-in-law, John Dryden, minister of agriculture. They were successful in persuading him to provide a budget for a tree nursery at OAC. The groundwork was being laid.

Although throughout his time away at school Zavitz tried to keep in touch with old friends in Ridgeway, he was not happy with the community's history. He observed how "the early settlers simply set fire to the bush as means of clearing land. Swamps were drained ...we must ... remember that it took our forefathers a long time to denude the land [so] that we had practically nothing left."[31]

Bertie Township was slow to take up Zavitz's ideas, but by the time he was at the peak of his influence in the 1950s, farmers did begin to fence their fields from cattle and undertake reforestation projects. Of these the most notable were the family of Bert Miller, for whom Fort Erie's local environmental group is now named. Despite the positive impact of Zavitz's ideas on Bertie's landscape, which like most of Southern Ontario is dominated by either his cousin's introduced soybeans or the forests that Edmund protected, Zavitz never enjoyed recognition as a hometown hero.

As with many visionaries, local celebration of Zavitz came after his death. In this case, the ceremony to honour Zavitz was indicative of how all three political parties of Ontario attempted to work with him in moving forward with his forest protection agenda. Although the ceremony was organized by the Conservative government of William Davis, the idea for it was put forward by the local Liberal opposition member Ray Haggerty. A few years later, Haggerty would work closely with the NDP legislator Mel Swart to curb urban sprawl through his newly launched Preservation of Agricultural Lands Society. At a ceremony on May 24, 1970, on the Fenian battlegrounds in Ridgeway, Zavitz was honoured as "the man who planted one billion trees." Fort Erie's mayor, John M. Teale, proclaimed a "tree planting and beautification week," in honour of Edmund Zavitz.[32]

Apart from his parents, Zavitz had few ties to Ridgeway after embarking on his high school studies in St. Catharines. One person he kept in touch with, however, was Edith Jackson, a director of the Bertie Township Historical Society. For twenty years she served as a telegraph operator at Queen's Park, before returning to Ridgeway. In her later years, she carefully built up a file on Zavitz's life's work. Her work gives a portrait of Edmund as a loving and concerned father to his three sons. In private correspondence, Edith Jackson expressed how Zavitz constantly "talked about his three sons" and how she always "heard about his family." He was also very "interested in the news from Ridgeway."[33]

Zavitz possessed an uncanny ability to use fishing trips to inspire a deep reverence for nature in those who joined with him. His extended family were the beneficiaries of these excursions, just like his political friends. His granddaughter Kathleen Mackenzie recalled, "We went in his Chrysler car on picnics many times and there was always a stream where we would catch trout, my introduction to fresh fish not canned. The lunch was always delicious and we sat on a plaid blanket under huge shade trees." He was also a loving grandfather. As she fondly remembered, "He was the best, always there with a lesson for me to learn, or a lap to sit on, or a special hug. I was truly blessed to know him."[34]

— Three —

Behind the Scenes

The distinction between a man-made forest fire and one created during a thunderstorm was responsible for some of the issues Edmund Zavitz faced. In nature, before the full force of the torrent is unleashed and a fire ignited through lightning, there are gentle rains that first douse the forest. Aboriginal people did not use fire rampantly as did the early pioneers in their efforts to clear the land, but rather used controlled burns as a technique of game management. As a result reserves such as those on Walpole Island and along the Grand River retained their extensive forests. Apart from Zavitz's hometown area around Fort Erie, where the local farmers did respond to his conservationist message during the 1950s, these reserves were the only parts of Carolinian Canada (largely southwestern Ontario) that met current watershed protection goals of 50 percent forest cover. Not one Native reserve in Ontario suffered the devastation of blow sand. One of the most successful Native leaders of nineteenth-century Ontario, the Six Nations Confederacy interpreter-mediator George Johnson (father of the poet E. Pauline Johnson), vigorously enforced laws against illegal tree cutting on the reserve. As a forest warden for their lands, he was attacked three times by poachers during his efforts to enforce these laws, but still he perservered.[1]

Although the Native people disagreed with excessive use of fire to clear forests, up until 1850 most people in Ontario who had the right to vote agreed that forests were barriers to progress and should be should be burned out quickly. For example, Elizabeth Simcoe, wife of Lieutenant Governor Simcoe, expressed a love of forest fires, likening them to spectacular fireworks displays in her diary. In the racist language of the times, these fires were jokingly called the farmers' "niggers," giving early pioneers deliverance from the hard work of clearing land. One of the biggest industries of the time came from the ashes of the burned-out trees — potash for soap and other industrial uses. Viewing deliberate forest fires as a wanton waste of resources, the influential historians Harold Innis and Arthur Lower supported Zavitz's efforts in warning against what they termed a "staples" economy whose ultimate product was a desert.[2]

Edmund's cousin, Harold Zavitz, eloquently decried the plague of burning forests for potash. The son of Charles Zavitz, he was following Edmund's example of becoming a professional forester. Harold termed the potash industry a "burnt offering of agriculture," since it required "sixty large-sized hard maples to make one large bag of potash."[3]

The forests of the fertile lands of Southern Ontario were largely cleared away by the 1850s. Settlers were forced to move north to acquire land and aspiring farmers began an assault on the rocky lands of the Canadian Shield. This photo shows such an effort in the Muskoka area, an example of what today would be called "rough pasture."

By the mid-nineteenth century, the dominant European-Canadian belief that forests were best burned for ashes began to break down. Farmers were suddenly finding it difficult to grow fruit in the harsher climate brought on by deforestation; their industry was on the verge of collapse. Moreover, by this time the arable land in Ontario had been largely cleared and farming was being moved onto unsuitable rocky soils on the Canadian Shield, causing yet more devastation to the fragile soil covering.[4] Watersheds suffered from spring torrents and summer droughts. Soil erosion was on the increase and some farmers were experiencing shortages of the wood needed to maintain their farming operations.[5]

Soon after their formation in 1862, the Ontario Fruit Growers Association took on the cause of reforestation. One of their strategies was to hold a competition for the best essay on "Shelter for Fruit Growing," which was won by John M. McAmish of southwestern Ontario. He lamented, "In years gone by the almost unbroken forest in which the country was covered formed sufficient shelter for fruit trees, but the necessity of sheltering them from the fierce sweeping winds is becoming greater every year, just in proportion [to the rate that] the country is being cleared up."[6]

A London-area recognized expert in scientific agriculture, William E. Saunders was a manufacturing chemist who became a leading originator of varieties of fruit and grain; one of his most important contributions was to work with his son, Charles, in the development of Marquis Wheat. A conservationist and public servant, William Saunders was appointed director of the new Experimental Farms Branch of the federal government's network of agricultural experimental stations. He created a massive shelter-belt of trees to protect crops from wind damage around the first experimental farm, which was established in Ottawa in 1886.[7] One of Saunder's employees, involved in this mission of demonstrating the value of trees in improving micro-climate for many agricultural crops, was James Fletcher, another pioneering Canadian environmentalist. He was a founding member of Canada's first environmental group: the Ottawa Field Naturalists.[8]

After Saunders became a full-time public servant, leadership of the Ontario Fruit Growers passed to D.W. Beadle of St. Catharines. In 1882 he served as the Ontario delegate to the American Forestry Congress meetings in Cincinnati and Montreal, accompanied by William Brown, the professor at OAC.[9] The American Forestry Congress served as an effective platform for the German-trained-and-born forester, Bernard Eduard Fernow, who was the first such professional to practise forestry in North America. Fernow would become an advocate of forest conservation within the American government as head of the U.S. Forestry Bureau, where he was eventually succeeded by his assistant, Gifford Pinchot. By wresting the control of forest reserves from the Department of the Interior, Pinchot was able to turn this position into the U.S. Forestry Service.

In a similar vein, Ontario's position of clerk of forestry, which began in 1889, was part of the Department of Agriculture. The first person to hold the post was Robert W. Phipps, but both Phipps and his successor Thomas Southworth were isolated voices and usually unheard. They wrote fine reports with great ideas for actions to protect forests, which were generally ignored by the Ontario government until Zavitz came on the scene in 1904. Southworth gave up in 1905. Although these clerks were not foresters, they did fulfill a role that Fernow called "talking foresters." B.E. Fernow came to Canada and became the first dean of forestry at the University of Toronto in 1907, with Edmund Zavitz among the faculty's first instructors that year.[10]

Rural journalists began to join in the protests against deforestation. One such remarkable figure was Thomas Beall, who wrote for *The Farmer's Advocate*. During the 1880s he was documenting the consequences of forest destruction on the Oak Ridges Moraine. His exposure of mill closures, summer drought that dried up creeks, and roadways blocked by heavy snowdrifts were also endorsed by Robert Phipps in his role as the clerk of forestry. He urged the development of a professional forest service based on the model from Imperial India, which, as mentioned earlier, outlawed grazing livestock in forests and over-large clear cuts that prevented natural forest regeneration.[11]

A tragic consequence of the "age of fire," when fire was routinely used to clear land for farms, mines, and townsites, was that ghost towns were created, the result of the desert-like conditions caused by the deforestation. This graveyard and a few ruins are all that remain of the village of Grant, now completely surrounded by one of Zavitz's greatest reforestation accomplishments: the 128,000-acre (52,800 hectares) Larose Forest, east of Ottawa.

With the support of conservationist logging interests led by William Little, a lumberman originally from Norfolk County but now living in Montreal, the provincial government took the first steps in permanently excluding agriculture from the remaining Crown lands of the Canadian Shield in Southern Ontario through the creation of the original Algonquin Park in 1893. The Algonquin Park Act was pushed through largely by the vision of the remarkable public servant Alexander Kirkwood, the chief clerk of land sales in the Crown Lands Division in Queen's Park, Toronto. The Act protected forests from agriculture and associated fires, which were threatening the headwaters of major rivers.

After the establishment in 1984 of Rondeau Provincial Park, a peninsula in southwestern Lake Erie, a new category of public land was established — the forest reserve. While similar to parks, forest reserves were protected from sale for agriculture and from tourists, who were then viewed as potential fire hazards. In dealing with Crown lands, Kirkwood was assisted by another public servant, Thomas Southworth, who became responsible for the establishment of forest reserves in 1896. One such reserve, now designated as Sleeping Giant Provincial Park, was established on the shores of Lake Superior; another was in Temagami. Similar in its focus on keeping agriculture out was the Eastern Reserve, a cut-and-burned-over area in central Hastings County, near Bancroft, in need of protection to encourage pine regeneration. Repeated burns had reduced the once-flourishing forest land to a moonscape of non-productive, bare rock.[12] However, despite these emergency pioneering conservation measures of the 1890s, they were not sufficient to rescue the provincial pine industry, which for years had been the dominant economic force in Ontario. The lack of pine regeneration in the burned-out forests would remain the primary threat for a number of years.[13]

The challenges facing a sustainable pine lumber industry in Ontario were not widely understood until 1912, when Edmund Zavitz was hired by the provincial government as chief forester of Ontario, and then director of the Forest Protection Branch. Even then, however, the province relied heavily on a federal-provincial entity, the Commission of Conservation that had been created by the Prime Minister Sir Wilfrid Laurier.[14] In general, Laurier was responding to pressure from the American government, then heavily influenced by Gifford Pinchot as an advisor to President Theodore Roosevelt, and by the Catholic Church in Quebec. The Commission, modelled on the example of Imperial India and the U.S. Forest Service, was the driving force in the province behind securing the regulation of public lands by professional foresters. The challenge of training foresters in Ontario was handed over to the newly created Faculty of the Forestry Department at the University of Toronto in 1907. Interestingly, it was one of Zavitz's first students, J.H. White, who would — behind the scenes — become responsible for much of the hard slogging in the forests, collecting information in preparation for the Trent Watershed report that would be published by the Commission of Conservation in 1913.[15]

The 1907–08 photograph of the staff and students of the Faculty of Forestry includes Edmund Zavitz (front row, far right) and his eventual long-time assistant, J.H. White (back row, second from the left). Zavitz would become the greatest champion of the University of Toronto School of Forestry, using his influence with the provincial government to hire its graduates despite great opposition from Northern Ontario farm and business interests.

When attending McMaster, Zavitz was fortunate to be at the right place at the right time. John Dryden, a founder of the Baptist University, encouraged his children to attend. He would have known all of the faculty, including Dr. Richard Smith, who had urged Edmund to become a forester. At the time, Dryden's position as minister of agriculture made him an administrator of the Ontario Agricultural College at Guelph, then affiliated with the University of Toronto. As noted earlier, Dryden, a successful and prize-winning breeder of pure-bred Shorthorn cattle, appreciated the university's agricultural goals. In keeping with the OAC faculty teachings, he wanted cattle kept in barns at night and the long-horned ones, more suited to doing battle with brush wolves ranging wild in unfenced woods, to be dehorned. He was aware of the problems of deforestation, being the member of Parliament from Ontario South, a riding that straddles the Oak Ridges Moraine.

Following his election as an MPP, Dryden had been appointed to serve on the Commission on Agriculture along with William Saunders. There he heard the testimony of D.W. Beadle, a fruit grower and author, who like William Saunders had been a president of the Ontario Fruit Growers Association in the 1880s. As a nurseryman in St. Catharines and a leading advocate of reforestation, he explained how his nursery and its biggest competitor, that of George Leslie of Toronto, couldn't grow White Pine economically. Other farmers told the Commission that such spruce were too expensive to purchase for them to embark on serious reforestation.[16] Other submissions detailed accounts of decreased orchard yields, the drying up of streams, the greater mortality of fall wheat from winter kill as a result of deforestation in Southern Ontario, and documented the impact of ruined forests from rampant fires on the Canadian Shield.[17]

After becoming minister of agriculture in 1890, Dryden learned more about the consequences of deforestation. He was in contact with such advocates of reforestation as William Brown of the Ontario Agricultural College and one of his students, Judson Clark. Dryden was involved at the cabinet table in the creation of provincial parks, forest reserves, and the employment of full-time game wardens. He was also in contact with Robert Phipps, the clerk of forestry, and his assistant, Thomas Southworth (who succeeded Phipps in 1895), both of whom wrote many reports calling for a professional forester service modelled on that in India, which were largely ignored at the time.

The final push that convinced Dryden of the significance of forests took place in 1903 during Zavitz's courtship with his daughter Jessie, who was then attending McMaster. The critical step was the formation of a Forestry Committee within the Experimental Union, the OAC alumni association founded by Charles Zavitz and dedicated to scientific agricultural research and experimentation. Dryden came to understand that it was most important to ensure Ontario agriculture's maintenance of high-quality standards for food. The importation of dairy products from Denmark into the Ontario market had been an awakening — these higher quality dairy products were vastly outselling those made in Ontario.[18] Education for farmers and OAC took on a new significance.

E.C. Drury, a founding member of the OAC Experimental Union's Forestry Committee, recognized the need for reforestation, having grown up in Simcoe County on his father's farm. Often he would have seen the night sky a blazing red in the vicinity of Midhurst and Orr Lake, illuminated by fires caused by the ruthless burning of forests to clear fields. He saw trains come into the Barrie Station "with the paint ... blistered from running through the massive forest fires to create farm fields on the Angus plains."[19] Drury later recalled that in Simcoe County there were pockets of land "so poor that it should never have been farmed ... but left in forest. There were also several areas that once had been covered with valuable stands of red and white pine. These areas had been cut over by lumbermen and the brush and waste timber being left on the ground to be burned over and over again until the seed in the ground was destroyed and natural regeneration was impossible."[20]

Another member of an Ontario political dynasty, Nelson Monteith, a close friend of E.C. Drury, joined the OAC Forestry Committee. At the time, he was a recently defeated Conservative MPP from the riding of Perth South, but would later become minister of agriculture. During his childhood, his father Samuel had turned their family farm in the Gore of Downie, near Stratford, into a conservation showcase by planting numerous trees. At the age of six, Nelson planted his first tree — a White Ash. By the year of his death, 1949, it was a majestic tree.[21] Roland Craig, another OAC graduate, had returned to Canada to resume a career with the Canadian Forestry Service after working briefly in the U.S. Forest Service. He served as secretary of the Forestry Committee and inspired its members by stories of what he was able to do in California for forest protection through the Forest Service, then led by Pinchot.[22]

Nelson Monteith, representative of the Perth South riding of Ontario (centre); with Arthur Herbert Richardson (left), originally hired by Zavitz as his publicist in charge of reforestation; and H. Crown. Photo circa 1948. Monteith played a critical role in persuading Zavitz's father-in-law, John Dryden, to create reforestation nurseries. Later, as minister of agriculture, he supervised Zavitz's work, and much later still, played a major role in the creation of the Upper Thames Conservation Authority.

Had it not been for Drury, Monteith, Craig, Charles Zavitz, and their friends in the Experimental Union, Edmund Zavitz's career as a forester may not have been possible, at least not so quickly. Although concrete evidence is elusive, it is likely that Zavitz was well aware of what was going on and may have been working with them in the background. From knowing Dryden personally, he might have had a good idea it was not an easy task to convert him to the reforestation cause. However, Dryden certainly knew Zavitz and had approved of his engagement to his daughter, so he must have been positively disposed towards him.

Pioneering environmentalists throughout Ontario also took part in the Forestry Committee's efforts to lobby the Ontario government. Another member, T.H. Mason, called on the Ontario government "to withdraw altogether from settlement these sections of the north country that are unsuited to agriculture, making such sections into forest reserves."[23] He was joined by J.H. Faul, a botanist with the University of Toronto. Faul stressed, "If the large area to the north of us, which is to conserve our water and timber supply, is to be properly cared for, it must be in the hands of foresters who have careful, scientific training."[24] Another important conservationist on the committee was Dr. W.H. Muldrew, a teacher from the Muskoka area and an author of an early book on Ontario forests. He obtained White Pine seedlings in a box from New York State and planted them at the high school grounds of Gravenhurst. He urged members to "look up the small plot containing the two hundred pine trees which my pupils set out. So far, the results have been very gratifying. The little trees took root and grew very rapidly, for it is well known that the pine will grow much more rapidly under cultivation than in the woods."

At the December 1902 membership meeting of the Experimental Union, on behalf of the Forestry Committee, Monteith moved, and Drury seconded, that

> the Experimental Union, recognizing the urgent necessity for taking action in the reforesting of the wastelands throughout Old Ontario, would recommend the Department of Crown Lands be requested to provide land to reforest areas sufficiently large to provide forest conditions in typical situations throughout Ontario, the Union undertaking to supervise the distribution.

A year after the resolution had passed, Craig explained to the Experimental Union that the Crown Lands Department had not responded positively. He did add, however, that Minister Dryden had given some "slight encouragement for the establishment of a nursery at the Ontario Agriculture property," for reforestation by farmers. Encouraged by Craig's report, the Experimental Union invited Dryden to attend its annual general meeting, where he heard pleas from Faul, Mason, Muldrew, and Drury for reforestation. E.C. Drury told Dryden:

> There are acres and acres of wasteland scattered among the farming land in some parts of Simcoe County, and in the other counties. It is not of much value agriculturally, but it was once covered with a good crop of tall pine. This land was cut over carelessly and burned, the means of reforestation were destroyed, and much it is now entirely waste. These areas could be reforested with great advantage, and our motion was that the Government should allow us to experiment with one such area, in the hope of inducing municipalities to take hold of the matter.

Dryden also heard from Charles Zavitz, who spoke of the need for a "permanent director" to lead reforestation and a "committee to assist him."

In his speech, Roland Craig outlined for Dryden what would soon become Zavitz's basic mission. He is recorded as having said, "At present it is almost impossible to procure plant material for forest planting, because many nurserymen rear seedlings for ornamental purposes, and expect fancy prices. It is, therefore, almost necessary that a nursery be established from which to supply farmers with plant material, free, or at nominal prices." He suggested that "two or three acres" be selected from the extensive grounds of the OAC campus.

Craig further explained to Dryden why a nursery was needed: seedlings could not just be dug up out of the woods. When they were "brought out to a light, exposed place," they were "unable to cope with the new conditions and [died]." Collecting seedlings in the woods also damaged their root systems. He stressed, "It is so difficult to pull them, and they grow so scattered in the woods that in the end, it is cheaper to grow them from seed in a nursery." To hammer home Craig's report,

Drury moved another resolution — that the province undertake "the practical re-forestation of areas sufficiently large to afford forest conditions." Craig's appeal to Dryden was strengthened by a lantern-slide presentation demonstrating the benefits of forest cover in Europe, especially for conserving soil and water for agriculture.

Minister Dryden, while receptive to the encouragement for him to expand and protect Ontario's forests, pointed out that the government was not yet prepared to act on their recommendations. He reminded them, "A great many men in public life are moved this way or that way by the politics in it, so that you have to keep up the agitation, and if you men who are just coming into usefulness in the country will keep stirring up this question, you will probably get what you are working for." He stressed that while he agreed with the views of the Experimental Union, its members had to be aware that those who wanted to protect forests were a minority in Ontario. He promised, however, that the government would "take action if public opinion showed itself sufficiently behind it."

In 1908, Zavitz made a photo documentation of sand dunes threatening to bury barns in Norfolk County. Despite evidence of such devastation, the Norfolk County Council of the time refused to use the legislation secured by Zavitz in 1911 for the development of county forests. It was not until the measure was heavily promoted by Premier E.C. Drury and encouraged by the publicity skills of Arthur Herbert Richardson that the council took action.

From Dryden's remarks, Drury understood that he was asking them to seek favourable publicity. Both men were sufficiently experienced in politics to understand how this could be done. Drury contacted newspapers that specialized in farm audiences across the province and that had long supported reforestation. The move generated the needed public coverage. Dryden, in turn, authorized the necessary funds for a reforestation nursery, simply putting the money in the Department of Agriculture's budget as part of the funding for the OAC. It was approved by the legislature.

It was this money that allowed Edmund Zavitz to set up a tree nursery on the grounds of OAC in the summer of 1904.[25] He accepted the $50 a month position offered by the OAC, and work commenced in the summer between the end of his first year of graduate studies in forestry at Yale and his attaining a MA degree at the University of Michigan.[26] The next year, 1905, he did a pilot planting of the seedlings grown in the nursery on the land that had been farmed by his grandfather on the ecologically devastated Oak Ridges Moraine. Some years later, in 1922, his uncle John Squair proudly observed, "The trees have thriven one sixteen-year white pine measuring 27 feet high and 6 and a half inches in diameter a foot above the ground."[27]

Zavitz decided to shift his forestry studies from Yale to the University of Michigan, largely because of the university's visionary dean of forestry, Filbert Roth, whom he had encountered sometime earlier at a meeting of the Canadian Forestry Association. Zavitz had enjoyed his studies at Yale, particularly "the field work in the Connecticut Hills," and was "beginning to notice what forestry meant in the economy of a country." The change of universities, however, was attributed to the similarity of Michigan's forests with those of Ontario, and especially to his having heard a passionate speech Roth gave in Toronto at the King Edward Hotel in 1904. Roth had described the denudation of the forests of Michigan, and asked, "Why do we pay millions to import our better grades of timber, and then burn up our forests where our timber should grow? When will the people realize that the legislature is ready to act as soon as the people demand it."[28]

Tragically, the response to Roth's question came in 1908, in the form of a catastrophic forest fire around the village of Metz in northern Michigan. The entire village went up in flames and twenty-five people died. Some two million square hectares of forests were burned, and the widespread devastation led to public protests and demands for major reforms. As a result, millions of hectares of those previously private lands, which had burnt up so terribly, were acquired for the extensive Michigan State Forests.[29]

However, as noted, the Metz disaster was on private lands, not the land that Roth had previously persuaded Michigan to acquire for state forests. During the time that Zavitz studied with Roth, the dean had reforested those lands through state nurseries. Appointed in 1903 to the University of Michigan, Roth acted quickly, and, in the spring of 1904, he supervised the planting of 50,000 trees. Before long, Zavitz would be establishing similar nurseries in Ontario.

Shortly before starting his work on the Guelph nursery, Zavitz had accompanied Roth to a Quebec City meeting of the Canadian Forestry Association. Here, he recalled that he had a "breakfast with three German foresters about spruce regeneration which was lost on me as they soon lapsed into German."[30] On the way home, Zavitz met William Little, who, despite his having extensive logging interests, turned out to hold similar views on forest conservation. It appeared that Little's opinions had been shaped by concerns over the role his own family's excessive logging in Norfolk County had played in creating the desert-like conditions there. During the return trip to Ontario, Little shared a quote from an old letter from his father with Roth and Zavitz. It was about a White Oak that had been cut, "the log squaring five feet, to be shipped to Great Britain."[31] This process of squaring logs for floating on rivers wasted a lot of wood.

Immediately after receiving his graduate degree from the University of Michigan, in May 1905 Zavitz was hired as a lecturer in forestry at OAC. This achievement, however, was tinged with disappointment. He had originally expected that his work in reforesting Southern Ontario would be done in co-operation with his friend Judson Clark, whose work involved protecting the forests of the northern Crown lands of the province. This was not to be. The hoped-for opportunity was blocked by powerful lobbying interests of Northern Ontario, championed by businessman and cabinet minister Frank Cochrane, and the advice of professional foresters was ignored. Ontario experienced ongoing environmental disasters, the aftermath of indiscriminate deforestation.

— Four —

Exiled to Agriculture, 1905–1911

EDMUND ZAVITZ WAS ABOUT to realize his childhood dream of working with nature. The challenge of working with trees to halt the spreading of Ontario's deserts lay ahead, but political changes were about to intervene. Powerful business interests caught up in the frenzied rush to exploit the riches of Northern Ontario were conspiring to eliminate the forest protection goals of his friend and mentor, Judson Clark. If their lobby was successful, much of the commercial forest of the north would be burned down to bare rock and Clark's proposals for forest protection would likely be ignored.

A provincial election in 1905 brought the Conservatives back into power and removed Dryden from his ministerial position. Clark clashed with the new minister of Lands, Forests and Mines, Francis "Frank" Cochrane (after whom the town of Cochrane is named). Having been in the role of opposition, Cochrane was very much a critic of the limited reforms, such as the establishment of more forest reserves that Clark pried out of the recently defeated Liberal government. Cochrane engineered the transfer of Clark's Forestry Bureau to the Department of Agriculture, thus shifting his duties to the promotion of agriculture in Northern Ontario. Clark, seeing this as a doomed venture hostile to his beliefs in conservationist forestry, resigned and moved to British Columbia, where he worked privately for forestry companies.[1] The ignoring of Judson's advice would lead to more catastrophic forest fires erupting in Northern Ontario, and the position of chief forester of Ontario would be vacant for seven years from 1905 to 1912, when Zavitz assumed the position.

Despite the brewing storm, Zavitz was about to embark upon a professional and personal honeymoon. Following graduation from the University of Michigan with his Master of Science in Forestry degree in the spring of 1905, he took on the position of lecturer at OAC. He was to "give lectures in forestry, develop forest nurseries to set up supplies for reforesting, attend Farmers' Institutes and begin a survey of the larger waste areas of Southern Ontario...."[2] In essence, what he taught at Guelph was designed for farmers, rather than future forestry administrators of provincial Crown lands. His role was to be supported by the new minister of agriculture, Nelson Monteith.

It was Monteith who put Zavitz in touch with his close friend, E.C. Drury. Their meeting in October 1905 was quite an arduous undertaking for Zavitz. He cycled the distance of about 160 kilometres. Given the shape of the roads then, it was necessary for him to pedal from Guelph to Toronto, where he likely stayed overnight, and then north to Drury's Crown Hill farm on the other side of Barrie the next day. The occasion was the start of a long friendship that would be sustained until 1968, the year they both died. At the time of their meeting, both men were settled into their respective careers, Drury as a farmer and Zavitz as a professional forester after a long academic training. Earlier, Drury had married the former village schoolteacher, Ella Partridge, after an engagement of several years. Zavitz was looking forward to marriage with Jessie Dryden in two months time, after a similar long engagement.

Zavitz stayed several days at the Drury farm. In his memoirs, Drury recalled, "I liked him at once and was very favourably impressed by him." They "very thoroughly" discussed reforestation challenges in Southern Ontario. The high point of the visit was their three-day horse-and-buggy expedition to view the marching deserts on the former pine plains of Angus and Midhurst in Simcoe County.[3] Zavitz described the Midhurst plain, then a very desolate place of about 2,000 hectares, as being "very thickly covered in stumps." The soil had been damaged by repeated forest fires, which were ignited by sparks from steam engines running on the adjacent Canadian Pacific Railway line.[4]

Amidst this desolation, Zavitz and Drury selected the future site of the Midhurst Reforestation Station. Zavitz later explained, "We walked across the field south of what is now the Station headquarters on Highway 26 and came upon a spring of fresh clean water bubbling out of a sandy bank. The stream wasn't very wide but it seemed to have a good strong flow. We knew there was much sandy loam in Simcoe which could be planted, and with the railway nearby this seemed to be the best location." The reforestation station later became the most northerly site of one of Zavitz's favourite trees, the Tulip Tree, which he had planted there. This species, also found in his favourite area, Norfolk County, is, like the White Pine, among the tallest growing trees in Ontario. The area, now Springwater Provincial Park, has become a haven for Black Bears.[5]

On the personal front, seven months after his OAC appointment, Edmund married Jessie Dryden on December 28, 1905. In December of the following year, the couple celebrated the birth of their first child. In an attempt to ease the pain of Dryden's electoral defeat, the couple named their first son John Dryden Zavitz in honour of his grandfather. More children followed. On March 14, 1909, their second son, Edmund Ross Zavitz, was born, and their third son, Dean Clarence Zavitz, would arrive three years later.

While Zavitz was strongly supported by Monteith, Judson Clark was having a tough time with Cochrane. This strain shows in the photographs that he and Zavitz were taking during their explorations, particularly in Norfolk County and the Lincoln County section of the Niagara Escarpment. No matter how healthy the forest around him, the chief forester of Ontario does not smile. During this period of fierce government opposition to forestry protection, the small number of persevering Ontario environmentalists closed ranks in support of one another. Zavitz would benefit from this allegiance when, in 1908, Cochrane and his assistant Aubrey White, the deputy minister, ordered Zavitz to cut the Tulip Trees in Rondeau Park, those Carolinian Forest

Two Billion Trees and Counting

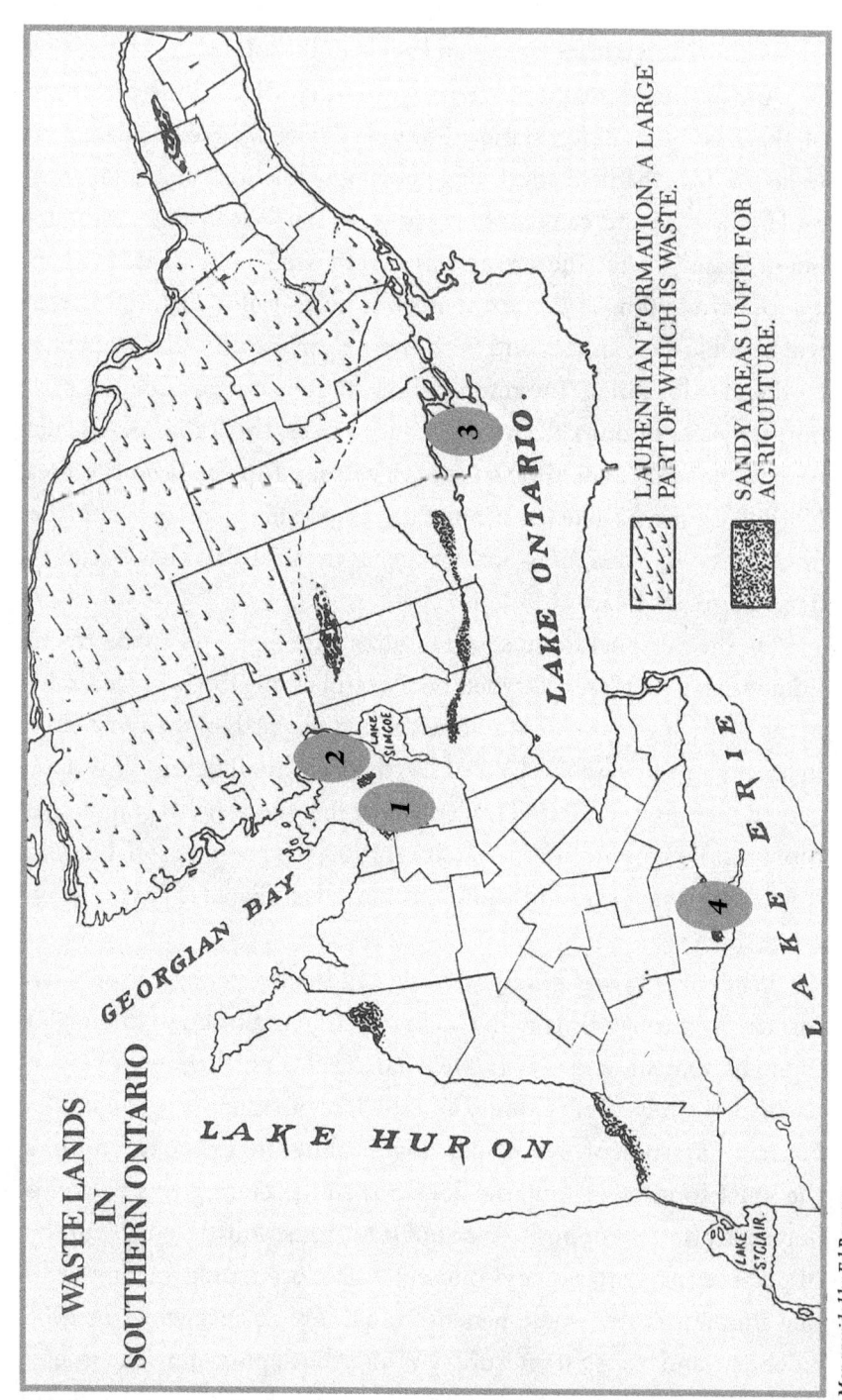

— *Exiled to Agriculture, 1905–1911* —

For three years after being hired as a lecturer by the Ontario Agricultural College in Guelph, Zavitz toured Ontario's desert-like wastelands, accompanied by E.C. Drury. The result was the 1908 report documenting the extent of the areas and providing suggestions for their reforestation. This map is from the wasteland report, complete with photos taken by Zavitz. The scene with the pines is near Angus, and the desert lands with the horse and buggy is on the site of the future Springwater Provincial Park, near Midhurst. The image near Lake Ontario shows cattle grazing, worsening the desert conditions in the future Sandbanks Provincial Park. The bottom photograph indicates an array of pine stumps in the sands in Norfolk County.

species he so revered. The trees were to be cut for commercial timber purposes, the proposal having been made at the instigation of the park superintendent, who wanted the revenues from logging for various park administrative purposes — the government encouraged

In 1908 Zavitz took this photograph of Black Ash and Shagbark Hickory trees in Rondeau Provincial Park, a peninsula jutting into the southern part of Lake Erie. With the help of Bernard E. Fernow and local residents, Zavitz was able to persuade the Ontario government to protect the giant old-growth trees found there. Today, logging is prohibited in Rondeau, as in all of Ontario's provincial parks with the exception of Algonquin.

parks to be self-financing. Aubrey White and Cochrane liked the idea and bullied Zavitz into marking the trees for cutting. Taking a stand, Zavitz complained to Cochrane that "cutting the timber in the park would result in a lot of opposition and criticism." In addition to the predictable local "storm of disapproval," Dean Fernow of the Faculty of Forestry came to Zavitz's defence, writing a letter to the Toronto *Globe* stipulating that the proposed cutting violated his belief in "esthetic forestry." The protests caused the Rondeau sentinels to survive.[6]

Although it was not part of his official duties, Zavitz would journey to Northern Ontario to photograph magnificent trees when time permitted. While in Algonquin he captured the spectacular Highland Grove of giant White Pines, later to be cut down, and documented the equally spectacular Hemlock and Yellow Birch trees also growing there. In 1907, Zavitz photographed a giant Aspen Poplar, almost half a metre in diameter, in Ontario's Clay Belt. Not knowing that this region would soon be burned out, Zavitz tried to send a conservationist

A rare self-photograph of Edmund Zavitz with a friend. This was taken at the corner of Abitibi and Moose Roads in the Clay Belt in 1907, when Zavitz was thirty-two years old. The tree is an Aspen Poplar with a diameter of about half a metre.

message by posting a copy of *Act to Protect Forests* on this particular tree.[7] The only official-capacity trip Zavitz took to the Near North, during the period from 1905 to 1912, was in the company of Monteith and Cochrane when they went to New Liskeard to select a site for an agriculture research station for the Clay Belt. Interestingly, the station became a fire break and, in 1922, helped save the town from a forest fire.[8]

Once elected, Frank Cochrane and his deputy Aubrey White refused to hire any foresters from the University of Toronto, or Guelph, or anywhere else. This was a particular blow to the University of Toronto, since this faculty had been created largely for the purpose of developing professional foresters to work for the Ontario government as administrators of Crown lands. The school's founders had wanted Judson Clark to serve as the first dean, but he, having personally experienced Cochrane's hostility to professional foresters, refused. As was his wont, Cochrane continued to have an army of politically appointed Crown Timber agents, many of whom were former loggers who owed their appointments to logging barons. At one time there were some 1,400 Crown timber agents. Their job was to administer northern forests, enforcing the province's forest laws on Crown lands, such as determining how much companies should pay government; they were even in charge of forest fire-fighting operations. (In time Zavitz would slowly reduce the significance of their role and replace them with university trained foresters.)

With the job market correspondingly reduced and the prevailing government attitude towards forestry a negative one, it was a challenge to build morale among forestry students. Zavitz, who by this time — 1907 — was also teaching a course on Dendrology (the study of trees) in addition to his OAC duties, did much to help students build pride in their chosen career and maintain a positive attitude towards the future. He was helped by one of his first students, James H. White, founder of the university's student forestry club. During his one year of teaching at the University of Toronto, Zavitz formed a close friendship with him. The two had a lot in common. Both were highly motivated to protect forests. Both had entered university as mature students in their late twenties (White having been a schoolteacher, Zavitz having worked at a variety of jobs for Eber Cutler in Ridgeway, before they attended university). Dean Bernard Fernow assisted in morale building by inviting students to his home for a Sunday meal, where he would enliven the evenings with his formidable skills on the piano.[9]

Fernow and Zavitz worked closely together nurture their forestry students and ensure their ideals remained after graduation by forming a professional organization of foresters, the Canadian Society of Forest Engineers. Founded in 1908, it sought "to cultivate an *esprit de corps* among members of the profession." Years later, in 1957, the Ontario chapter persuaded the province to create the Ontario Professional Foresters Association as a registration body. In 2000, it became a licensing and regulatory body, established to protect the public from unprofessional practices.[10]

Although initially Zavitz only had few employees to help with the nursery, planting trees on the Guelph campus, shipping seeds to farmers, and so on, he did build a strong professional spirit among his staff. After the OAC Guelph nursery was moved to St. Williams in 1908 (it was his idea and he had to work hard to persuade the government to fund this move) it remained an OAC nursery until 1912, when it was transferred along with Zavitz, to the Department of Lands and Forests. Zavitz's staff then took on the task of reclaiming the deserts around them by planting trees, proudly calling themselves "pioneers of sand." One of these men, George Lane, a faithful and dedicated employee, passed this sense of dedication on to his son, George Ritchie Lane. With degrees in both agrology and forestry, George Jr. would become superintendent of the Midhurst Forestry Station and be assigned to the task of reforesting the sand wastes of Simcoe County in 1922.[11]

The University of Toronto forestry dean, Clifton Durant "C.D." Howe, Fernow's successor, explained why strong morale among Zavitz's employees was so important, when he persuaded the government not to implement a scheme to use prison labour for government tree-planting. Howe sent a letter, dated January 4, 1928, to the then minister of lands and forests William Finlayson, pointing out that

> the raising of trees economically requires special training. It cannot be done well or on a casual or short term basis. Frank Newman [hired by Zavitz as superintendent of St. Williams], has materially reduced the cost of output from year to year because he has trained his labour. The more experience a man has the more efficiently he can work and the more efficiently he can raise trees. If we are going into the project to raise trees cheaply and economically on a large scale, the quality of the labour should be seriously considered.

Howe also pointed out that the use of prison labour to plant trees failed in the only place where it was tried — New Zealand.[12]

Prior to the First World War, Zavitz was able to make 400,000 trees available to Ontario farmers for free annual distribution under his co-operative planting program. The program started when Zavitz was at OAC, began with 10,000 trees, and ran in various forms from 1905 to 1996, when this type of distribution for free from government nurseries was discontinued.

Zavitz reported on the first year's effort of pilot planting projects to the 1906 meeting of the Experimental Union.[13] In his address, he was able to provide a detailed account of the tree planting on his grandfather's former farm on the Oak Ridges Moraine. He explained:

> The field was originally cultivated and cropped, but to-day all the knolls and more exposed portions are drifting sand. The soil produced splendid growth originally. Upon turning over a few inches of the surface sand there is plenty of moisture, which makes it even a good tree soil as the roots extend down into the moist portion. Portions of the planting area were covered with sparse growth of grass and weeds. The planting material used was Norway Spruce and native White

— *Exiled to Agriculture, 1905–1911* —

In 1905 Zavitz planted pine seedlings on the farm owned by his maternal grandfather, Edmund Prout, then photographed the results of his work. Note that the seedlings are planted in desert-like sand.

Francis Squair is shown with the pine seedlings, which are now four years old.

The son (name not known) of Francis Squair is shown in 1985, standing amid the pine forest planted by Edmund Zavitz in 1905.

Pine. The owner had collected about a bushel and a half of Red Oak acorns, which were also used on the better portions of the area. The evergreens were taken from the express office to the source of the scene of planting and heeled in by a small stream in a protected place. In the portions of the field, which were covered with grass and weeds, a very light furrow was skimmed out every five feet. This furrow answered two purposes. It made a planting line to follow, and more important, it gave the planting spot a better moisture content. The plant, being placed on the side nearest the grown-out sod, is protected the first year from being overgrown by the grass and weeds. It was afterwards found that in many parts of the area the furrows gathered leaves and other drifting material which formed a mulch for the plants.

The actual planting was done as follows. Planting holes were made by driving in the spade and moving it back and forth. The plants were inserted and the roots firmly covered by a boy who carried plants in a twelve quart pail half filled with muddy water in which the roots of the plants were submerged. Plants were eight to twelve inches high with very compact root systems. From seventy-five to one hundred plants could be placed in the pail at once. Plants were spaced about five feet apart, which would mean that an acre required 1,742 plants. The Red Oak acorns were dibbled in with a sharp stick or dibble and spaced as the same as the trees. The actual work required about three men per day per acre for the White Pine and Norway Spruce. The acorns were dibbered [planted using a hand-held dibber] at a much faster rate, a boy planting an acre per day.[14]

Zavitz went on to claim the pilot planting a success. He found that about 95 percent of the White Pine had grown well. The Norway Spruce, however, had given poorer results, with only about 75 percent living. The White Pine, he stipulated, had passed the most dangerous period of growth and the land was now well-stocked with young trees. He also reported that another planting on this site in the spring of 1906 was doing well.

The other pilot planting that Zavitz was supervising in Perth County (possibly at a site suggested by Monteith as he was from that area) was also described:

> This planting differed from the sand-land planting chiefly in the preparation of the planting holes. The hillside was too steep to put horses in so no furrows were run. The planting holes were about five feet apart each way. In making these spots a piece of turf was cut out from the side of the hill with a grub hoe or adze and the sod loosened and freed [of] stones with a pick and grub hoe. After the soil was loosened the plant was put in with a strong spoke as in the sand-land planting, although some care had to be exercised to cover the roots.
>
> The planting took about twice as much labour as the previous one owing to the time spent on preparing the planting holes. Frequently in hill-side planting, it is possible to run furrows which can be used in planting lines. These furrows should always be run parallel along the hillside rather than up and down in which case washing down the furrow might occur.[15]

In response to a question about weeds interfering with reforestation, Zavitz made one of his quips — "I think you will have very little trouble on most wasteland," due to the fact that such barrens were largely sand, devoid of grass. As for the actual sand dunes, he warned against heavy cultivation on these hills. In his experience, cultivation should only be done on heavier soil. Even here, he cautioned, it was better to cut weeds out by hand.

Zavitz photographed the planting of White Pine at Guelph in 1905. Today, these pines are known as Zavitz's Pines.

As part of his teaching role, Zavitz initiated some reforestation projects on the OAC campus grounds. The entrance to the university's renowned Arboretum is now graced by the "Zavitz Pines." He also undertook reforestation in Guelph around the city's wellhead protection zone for the springs providing its drinking water. Over the next four decades, he would continue to establish wellhead plantings to protect sources of drinking water from contamination. Trees, which help purify polluted runoff and thus eliminate contamination of shallow aquifers, were planted in Beeton, Hanover, St. Thomas, Orangeville, and Midland.[16]

While at OAC, Zavitz produced his "Farm Forestry" booklet, which sought to develop appreciation among farmers for the forest "community," or "organism," where the forest acts as a living being in itself and the living creatures within it all work together. He warned of the dangers of lowered humus content of soils. Humus, formed by the decay of organic matter, was essential to soil productivity, and as he pointed out, humus "makes heavy soils less stiff and sandy soils more binding." It provides nutrients for trees and increases "capacity for absorbing water."[17] Declining humus content was causing Ontario forests to die, Zavitz warned, as was exposure to wind. Gusts in poorly protected forests dispersed leaf litter, preventing decay into forest soil, meaning seeds would lack sufficient protection to germinate.

Zavitz urged farmers to see their forests as a "permanent resource" and a "necessary part of the farm economy." Aware from his youth of the antics of local lumber moguls like Eber Cutler, Zavitz described sawmill owners in Southern Ontario (and Cutler was one of them) in villainous terms. Sawmill owners like Cutler did not own huge tracts of land but often tricked farmers in their area who were hard up for money or less educated to sell their trees, which were stripped, thus creating wastelands for small sums. Conservation authority reports in 1940s and 1950s are full of the same type of problems. One of the major goals of associations like the Ontario Forestry Association and Ontario Woodlot Association is to protect landowners from this sort of villainy. Zavitz urged that logging be done with greater care. Logging barons, in his experience, simply "[took] out the best timber, leaving only slash." He repeated warnings against livestock grazing in woods and urged that clumps of trees in fields be used to shelter animals from the heat of the sun.[18]

Zavitz's studies of the Oak Ridges Moraine during his time of teaching at OAC, although shaped by the stories he learned in childhood, were also heavily influenced by his uncle John and cousin Francis Squair. His 1908 "Report on Reforestation of Wastelands in Southern Ontario," prepared for the Ontario Department of Agriculture, provided the first detailed map of this desert-prone moraine. It warned that streams flowing from the moraine had "almost ceased to exist, except for a short time during spring freshets" because of the extensive deforestation. He stressed the importance of streams flowing freely from

Edmund Zavitz took this photo of encroaching desert sand, which was in the process of burying a road on the Oak Ridges Moraine near Uxbridge in Ontario County (now Durham Region). His 1908 wasteland report, prepared for the provincial government, drew attention to the need to reforest the area.

the moraine "to the agricultural country through which they flow" and their being "important to many areas in this district as a future source of fresh water supply." Zavitz urged that the Oak Ridges Moraine be reforested to safeguard and improve book trout populations. He believed that "their value simply as trout streams" flowing from the Oak Ridges Moraine "should prove a strong argument for protecting their sources." He estimated that 75 percent of the Oak Ridges Moraine was "wholly unfit for farming" and therefore should be reforested. As a priority, he urged that action be taken on 2,023 hectares of wasteland in Northumberland County's Haldimand Tract and the 10th and 9th concession of Clarke Township in Durham County.[19]

Zavitz was on his own in documenting the wastelands around Lake Huron, where he lacked friends and relatives who could act as guides. Here, however, in view of the subsequent decision to modify reforestation by cutting some trees in Pinery Provincial Park to perpetuate a rare savannah ecosystem, the cautionary words of his 1908 wasteland report are quite prophetic. He pointed out, "Much of the Lambton County area presents a problem of protection rather than one of reforesting." His 1908 report gives an example of how protecting this area from abusive human activities should encourage natural forest regeneration. He photographed an "old scrubby red pine," which his report stressed "is rapidly restocking the surrounding ground with young growth." Another challenge facing Zavitz then, and also later public foresters, was that the role of Native land management methods in perpetuating savannah ecosystems was largely unknown.[20]

When Zavitz acquired a cottage at Turkey Point in 1908, he became involved in an intensive study of the wastelands of Norfolk County. He was contacted by the owners of the McCall Furniture Factory in St. Williams, which specialized in the manufacture of church pews, prayer desks, altars and communion rails. The McCall family saw his arrival as providential. Afraid of losing their business because of the loss of trees, they wrote and invited him to visit. Zavitz cycled the several kilometres to their home in St. Williams and met both Walter McCall and the Conservative MPP for Norfolk South, Arthur Pratt. The three then proceeded to undertake a study of the local expanding deserts.[21]

Zavitz reported on the extent of Norfolk's deserts of shifting sand in his 1908 wasteland report. He found that there were 4,040 hectares

This photograph of wastelands in Norfolk County, showing a former pine forest quickly turning into sand piles, was taken by Zavitz in 1912.

of what he termed "shifting sand," which he urged "must be placed under forest management." He also warned that in Charlottesville Township there were another "several hundred acres," which were "gradually developing into sand dune formations." His report contained spectacular photos of the tangled roots of White Pine stumps sitting in the spreading sands.[22] Zavitz made a strong economic case for reforestation in Norfolk. In 1908, the crops being grown there, notably rye and buckwheat, could barely pay the farmers' expenses. Restoring White Pine would be a better investment in the long term. Interestingly, the methods he used in calculating the cost of forest establishment, potential timber yields, annual management costs per acre, and financial returns are still employed today.[23]

Following Nelson Monteith's narrow defeat in Perth South in the 1908 election, Norfolk's Arthur Pratt provided crucial parliamentary support for Zavitz's reforestation efforts. A schoolteacher before being a legislator, Pratt spoke with great passion in the provincial assembly on the need to restore Norfolk's forests. He later recalled, "When the budget debate got underway in 1908, I gave the House some history of the days of Norfolk, of the great White Pine that the early settlers found, of the tall trees from Norfolk that were masts for the British navy."[24]

Frank Cochrane, however, who was very skeptical of claims made by the forestry profession, could not be convinced by Pratt's eloquence alone. The minister, who had earlier clashed with Judson Clark, had to see it to believe it. In response, Pratt organized a tour that would bring Zavitz and Cochrane together right on the Norfolk County's ever-marching sands. Cochrane was astonished; he had never seen anything like this devastation in his home Nipissing East riding. In Norfolk he saw great blow pits of up to three metres deep dug out of the sandy soil. The wind had shot up sand dunes two-and-a-half feet tall. Farms were menaced by waves of sand being blown in by great winds. On a particular windy day, dark clouds of sand eerily covered the horizon, appearing like a threatening rain storm. Fence rows were buried in sand, frequently appearing as lines of stumps and dead trees, scary memorials to the living green ones of the past.

Having come face-to-face with such devastation, Cochrane became convinced — something needed to be done. Shortly afterwards, when he and Pratt returned to Queen's Park, Cochrane told him that the steps needed to move the Guelph Reforestation Station to St. Williams would be swift. An order in council was passed, which provided $5,000 for land options and expenses to transfer Zavitz's nursery that same year, 1908.[25]

Zavitz's photograph, taken in 1912, shows an unidentified man beside a giant White Pine stump with the four-year-old pine trees growing in the sand. This is the beginning of the extensive forest Zavitz had planted around the St. Williams Forestry Station in Norfolk County in 1908.

Over the years between 1908 and 1912, Zavitz assembled the St. Williams Station within wastelands to be reforested and blocks of healthy forest to be protected. During this period some 1,080 hectares from thirteen adjacent properties were acquired for the purpose. A water tower was constructed for both fire protection — a high priority — and the watering of seedlings. The system worked well; there would not be a major fire until 1933.[26] Zavitz built a careful case to persuade the government to reforest a large acreage. He wanted to acquire enough land to require a resident custodian to manage it as a full time job. The same figure of 1,000 acres, today's 404 hectares, was later used for the Agreement Forest Program; all of these forests would have guardians who lived in a home surrounded by the trees that they cared for on a full-time basis.

The first hundred acres, acquired in 1908, consisted of a block of sand where 350,000 White and Red Pine seedlings were planted immediately. As the plantation grew under careful management, it assumed the appearance of a magnificent old-growth mixed Carolinian forest. Over time, deciduous species such as Sassafras, oak, and maple mixed in with the pine through naturally dispersed seeds. The pines now tower 27 metres tall. Author Harry Barrett remarked, "The straight, clear trunks of those majestic pine trees with their healthy, lofty canopy of green needles, give some indication of what the forest of the Norfolk sand plain must have been like when the occasional Attawandaron hunting party passed along the lofty trails beneath the canopy of these forest giants in the early 1700s."[27]

The process of resurrecting the magnificence of the once great forests from shifting sand dunes took considerable time. Zavitz had seedlings planted on the sand "closely together enough so the ground [would] be well-stocked to prevent the growth of grass and weed seeds." The mass of seedlings also protected the ground and formed "the needed humus in a shorter time." Once the soil had improved, "a carpet of wildflowers appeared, including lady's slipper, wild calla lily, turtlehead, fringed polygala, and the rare devil's bit."[28]

Zavitz monitored the ecology of the restored St. Williams Forest in the company of dedicated naturalists, such as the dairy farmer and botanist Monroe Landon, OAC professor Dr. Fred Montgomery, and local naturalist Harry Barrett. He also engaged in intense historical research on past ecological conditions. French explorers and missionaries François Dollier de Casson (a Sulpician priest) and Father René de Bréhant de Galinée found themselves having to winter over 1669 to 1670 on the north shore of Lake Erie in the vicinity of today's Port

Dover. While there, they recorded information in the local flora and fauna, which would became part of the Sulpician Journals. Zavitz drew on these seventeenth-century accounts and discovered that the local Carolinian ecosystem area then had "immense herds of deer, game of various kinds and the abundance of trees, wild grapes, and apples."[29]

While Zavitz developed a close circle of like-minded friends in Norfolk County, his views were not shared by its local councillors until much later in the 1920s, even though his 1908 wasteland report had provided evidence that township councils could, through a modest increase in property taxes, finance a program of public land acquisition, forest protection, and reforestation. He had made these calculations based on the cost of reforesting the sand dunes of the worst-hit municipality in the province, South Walsingham Township in Norfolk County. Zavitz estimated that its "10,000 acres of sand dunes" could be acquired and reforested with a one-mill increase of assessment of its current 16-mill rate.[30]

On the basis of Zavitz's report, and reinforced by resolutions of the counties that straddled the Oak Ridges Moraine, the Ontario government passed the County Reforestation Act of 1911. The Act clarified that reforestation, of the sort Zavitz had urged in his 1908 wasteland report, was a legitimate undertaking for county government. Despite all of the evidence, not one municipality in the province would use its reforestation powers until E.C. Drury became premier, and, with Zavitz's help, aggressively promoted it.

Interestingly, the greatest interest in using the County Reforestation Act came out of the rocky waste deserts of the Canadian Shield, where forests had been destroyed by fire, farming, and excessive logging. Zavitz, although he did not go into great detail, considered the entire area where farming had been attempted on the Canadian Shield as a wasteland suitable for reforestation, a view that was reinforced by C.D. Howe and J.H. White in the 1913 Trent Watershed Survey. Hastings County did consider the purchase of 809 hectares (north of the Tweed area) of the sort of wasteland identified in White's report as being available for $200, but nothing came of it.[31] Seemingly, opponents of reforesting the wastelands of the Canadian Shield expressed outrage at the way poor people living on the Canadian Shield were described, and the earlier initiatives involving use of County Reforestation Act were reversed.

While Zavitz's ideas were not embraced by many councillors, he did build up support among farmers. Mr. Goth, a farmer and OAC graduate, while speaking at an annual meeting of the Experimental Union, said:

> I turned the cattle out of the woods and planted trees, and now have trees reaching from 19 inches to 4 and 5 feet. Two years ago, I started under the direction of Mr. Zavitz to plant trees on the hillside and last year had very good success. Our success [has] come from simply following instructions. Land and hillsides that are now eyesores can be made beautiful by planting them in trees. I often wonder when I see the hillsides naked and bare why the [trees] were ever cut down; the [lands] are of no value for cultivation. We have these hills all about us, and we have started to wait because I am not going to see all that is in the world before I go. I am going to leave something behind, and I would just as soon leave some pine trees growing as a bank account. If you do not succeed the first time, plant again. Mr. Zavitz will help you along with the work, and if you are quite green in the work, he is not afraid to take off his own coat and help you do it.[32]

Zavitz's role as an OAC lecturer suited him well since it combined both teaching and public service. Many of the students he taught were inspired by his conservationist message and became colleagues in the cause. Two of them, John Bracken and Walter Jones, became long-time provincial premiers, in Manitoba and Prince Edward Island, respectively. Another former student, Watson Porter, worked closely with Zavitz to create conservation authorities. Zavitz had a major influence on another remarkable conservationist, H.R. MacMillan, also a former student, who as the chief forester of British Columbia from 1912 to 1919 created the British Columbia Forest Service. Later, as a businessman he developed the MacMillan-Blodel Corporation, which became one of the largest forestry corporations in Canada. Known as "H.R.," he actively supported environmental causes, including the creation of more provincial parks. One of the parks bears his name: MacMillan Provincial Park, which contains the great Cathedral Grove of towering Douglas Fir trees. A letter from MacMillan, written in April 1966 just after Zavitz entered a Brampton nursing home, is illustrative of the caring interest Zavitz took in his OAC students. MacMillan credited his teacher with the wise guidance that proved to be a "turning point" in his life, advice that had directed MacMillan to employment with the Canadian Forest Service.[33]

Zavitz's spectacular large-scale reforestation efforts around the St. Williams Station contributed to his growing, favourable national reputation. The ensuing positive publicity caused a powerful lumber baron from the Ottawa area, Senator William Edwards, who owned the land that today is 24 Sussex Drive — the official residence of the Canadian prime minister — to approach Zavitz. Edwards was considering a reforestation project on his property in Rockland. Although the planting was successful and became an important demonstration site to advance reforestation in Eastern Ontario, most of the forest was later razed for an arena parking lot.[34]

Zavitz's growing national reputation was also an asset in a series of events in 1911 that ultimately would end the exile of forestry to the Ontario agriculture portfolio. His impressive reputation intersected with protests in both Ottawa and Quebec regarding Ontario's refusal to allow foresters to manage northern public lands, a decision that was resulting in widespread fire, waste, and destruction. The final blow was a calamitous forest fire in Northern Ontario — the Great Porcupine Fire of 1911. Seventy-five people died in this inferno that consumed

2,022,787 hectares of forest. The towns of Cochrane, South Porcupine, and Pottsville were completely burned to the ground. Some buildings in Golden City (later named Porcupine) and Porquois Junction were damaged. Three thousand people became homeless. The largest gold mines in Canada, Hollinger and Dome, were incinerated along with eight other working mines.[35]

The outrage stemming from the fire, as expressed in Ottawa and Quebec City, focused on exposing the remarkable difference in forest management on public lands that existed between the two provinces. With the support of the Catholic Church, Quebec had developed the same type of management of public lands by professional foresters that was the norm for the British Empire, whereas Ontario's policy had negated the employment of such trained foresters. The leadership for Quebec's approach to forestry management had come from priests at the University of Laval, especially its rector, Monseigneur J.C.K. Laflamme, and Monseigneur Charles P. Choquette, who served as one of Quebec's representatives on the Commission of Conservation. They had persuaded the Quebec government to embrace scientific forestry and pay to send Joseph Bedard and C.C. Piche to be trained as foresters at Yale. After completing his studies, Bedard became the first director of the Quebec Forest Service, which imposed tough regulations compelling the cleaning up of logging slash. The Service also created a permit system for farmers seeking to clear land with fire, and imposed jail terms for violators — in one case for two years. The message to the public, of the need to protect forests, was spread by the church through two circular letters read out in every parish in Quebec. Bishop Roy of the Diocese of Quebec proclaimed the need for a "battalion of forest missionaries" to protect forests with "the zeal of an apostle and patriot."[36]

The horror of the Porcupine fire dramatically illustrated the benefits of the Quebec Forest Service, as opposed to Ontario's persistent adherence to logging and farming interests that stubbornly opposed the forest-protection model, which by then was widespread in both the Empire and the United States. Both sides of the provincial border were hot and dry that summer of 1911. While fires did erupt in Quebec, the efficient work of the Quebec Forest Service suppressed them without the loss of human life or homes. But now Quebec was facing the danger of forest fires that had originated in Ontario and migrated into their territory.[37] Railways were seen as a major culprit in the fires that erupted into the Porcupine disaster — the escape of hot ash from pans, and cinders flying from the steam engine stack, could easily ignite a burn, especially during a hot, dry season.

In the long run, as will be seen, Zavitz would benefit from the hard-slogging investigative work of J.H. White when he was secretary of the Commission of Conservation's Forestry Committee. While Zavitz was spending most of his time in Southern Ontario, between 1907 and 1912 White was photo-documenting areas of concern in the forests of the Canadian Shield. In 1912 he persuaded the Commission of Conservation to publish his dramatic photographs of rail lines passing through bare rock, scarred by skeletons of fire-charred trees, in its annual reports. The soil had gradually lost its fertility, the result of repeated fires destroying its humus content. White had documented that all along the rail lines in Northern Ontario the land surrounding them had "been burned at one time or another for the entire distance of 500 miles." He added, "Not much has escaped except for spruce swamps ... in a vast extent the country is burned so repeatedly that there is nothing but bare rocks."[38]

Something had to be done to halt the ongoing threat of fires in Northern Ontario forests. The federal government put pressure on William Hearst (the replacement for Frank Cochrane, who had become a federal member of parliament) to employ Zavitz, who by this time had built up considerable popularity among Southern Ontario farmers. The director of the Canadian Forest Service, R.H. Campbell, had made an attractive job offer to Zavitz to establish a tree nursery in Saskatchewan, thus creating the possibility that Ontario might lose its top forester. At the same time, the federal government had passed legislation requiring provinces to appoint foresters to regulate the railways. Quebec would be the first to act on the legislation.[39] In Ontario, the difficult task of persuading Minister Hearst to comply with the federal legislation fell on Aubrey White — a challenge for the deputy minister. Following the Conservatives' rise to power in 1905, White had been the enforcer of their policy that foresters would have no role in the management of public lands, but rather that power would remain with 1,400 Crown timber agents. Now he was to see that this policy was reversed.

Aubrey White is reputed to have informed Hearst of his admiration for the way Quebec, with the help of the Catholic Church, had cut down its loss of forests from fires, and he expressed the opinion that Ontario clergy would be pleased to give sermons on this topic as had already happened in the United States. White added that "the higher officials of the various railways" themselves were giving "reasonable support" to the new federal regulations. Their previous worthy effort, "without provincial inspection," White warned, would be doomed to failure. He also reassured Hearst that "the work of the railway Inspectors is ... closely allied to that of general fire protection."[40]

As a result of the pressure from Ottawa and Quebec City, on November 8, 1912, Zavitz was appointed chief forester of Ontario. As had been seen earlier with Judson Clark, however, this was an honorific title rather than a role with significant authority. His real power came from his simultaneous appointment as director of the newly created Forest Protection Branch of the Department of Lands, Forests and Mines. Initially, the Forest Protection Branch under Zavitz's direction had five employees — a distinctive group in a department dominated by former lumberjacks who were hostile to his conservationist agenda.

Zavitz was able to transfer his position at St. Williams on very favourable terms. Although his teaching at Guelph had ended, the reforestation work he did at the college from 1908 to 1912 had become a St. Williams campus for OAC, and remained under his control in the new Forest Protection Branch. In addition, he received new responsibilities for protecting forests from fire caused by railways. Zavitz accentuated the differences between former lumberjacks and forestry professionals within Lands and Forests by giving his employees their own uniforms and distinctively marked vehicles. This was in keeping with the role of foresters throughout the British Empire and in the U.S. Forest Service,

Two Billion Trees and Counting

The pines in this 1975 photo of the Rockland Tree Plantation are sixty-one years old. The three foresters whose faces can be seen are, (l-r): R. Philip Anslow, Wim Vonk, and Clarence Coons. Much of this forest was razed after the ice storm of 1998 to accommodate an arena and a parking lot.

where badges of authority were important in clashing with dangerous, frequently armed opponents. While Zavitz had been largely isolated from effective conservation work in 1905, he would ultimately alter the entire pattern of the Department of Lands and Forests to become an effective instrument of conservationist, scientific forest management.

The Province of Ontario, however, had only complied with federal law in a very minimalist fashion. The big loophole in its compliance was to exempt provincially chartered railways from its regulations — a dangerous loophole, as it was the provincially chartered railway, the Temiscaming and Northern Ontario (T&NO) that went through the heart of the most fire-prone part of the province, the Clay Belt. Moreover, the threat of fires from railways was increasing as newcomer farmers were ruthlessly burning forests to clear land for cultivation.[41] Zavitz was facing an enormous challenge as he entered his new role in 1912.

— Five —

The Struggle Against Indifference

EDMUND ZAVITZ'S APPOINTMENT AS chief forester of Ontario in 1912 was a remarkable victory for forest conservationists across Canada. Finally, it was believed, the dangers of burning the country's forest down to bare rock and the need for reforestation in general would receive real attention. However, they also understood the very limited nature of their victory, only achieved through some arm-wrenching pressure on the Sir James Whitney's Conservative government. The province's reluctance to act was tempered even further by its decision to exempt its own chartered railways from the national system of regulation — to be monitored by foresters. E.C. Drury, now a frequent speaker at meetings of Farmers' Institutes around the province, had made reforestation a favourite topic; he saw Zavitz's new role as an "answer to prayer,"[1] though his allies would have to do a lot of praying after 1912 to get such a project underway. Forced on the Whitney government by outside pressures, Zavitz would have very slim resources for protecting Ontario's forests from fire.

The challenges facing Zavitz were summed up well in a February 17, 1917, memorandum from the deputy minister of lands, forests, and mines, Albert Grigg, to his minister, Howard Ferguson. Grigg, whose personal honesty was never doubted despite his being in the midst of timber scandals of the time, attempted to do his best to support Zavitz. Following the catastrophic Matheson fire of 1916, Grigg told Ferguson that Zavitz needed more resources to do his job. Grigg explained that it was

> utterly impossible to make any intelligent headway with reference to the re-organization of our fire-fighting scheme for the Province under existing conditions. If Mr. Zavitz is to make any headway it will be necessary that he be given a competent assistant at once and further that some adequate office accommodation be provided. At the moment the only office staff that Mr. Zavitz has is one stenographer with her desk located in one of the hallways of the North Wing. The office, which he himself is occupying, is altogether inadequate for the work he has in hand ... the correspondence is piling up in the Department and the work is practically at a standstill.[2]

This apple orchard in Prince Edward County being buried in sand is another example of the devastation caused by the deforestation practices of early nineteenth-century Ontario. The area was reforested under Zavitz's supervision. In 1960, it became Sandbanks Provincial Park.

While focused on the need to control northern forest fires, Zavitz continued to supervise those government programs for reforestation that had been transferred with him from the Agriculture Department to his newly formed Forestry Branch. Apart from his work around St. Williams, the problems of drifting sand were accelerating. One of the most serious areas was in Prince Edward County on the shores of Lake Ontario. Here, the drifting sand dunes were burying some of the most fertile farmland in the province. Apple orchards were being entombed by marching sand dunes, creating scenes similar to the Sahara. In his report as chief forester in 1917–18, Zavitz warned that in the county, "The formation is made up of sand ridges with very little vegetation left. These ridges are shifting, forming dunes, which will be more difficult to control than the blowing sand in level areas."[3]

Until Drury became premier in 1919, Zavitz did not hold much hope for measures to extend the reforestation regulations needed to maintain the fertile land of Southern Ontario. With the onset of the First World War, the programs were scaled back even further, and the scarcity of labour and supplies led to Zavitz's co-operative planting program for farmers to plunge from 350,000 to 40,000 trees planted per year.

This decline was an unfortunate beginning for Frank Newman of Merrickville, Ontario, who was a University of Toronto forestry graduate. His appointment as the first superintendent of the St. Williams Reforestation Station was the first time, apart from Zavitz himself, that a professional forester had been hired by the Department of Lands, Forests and Mines since the departure of Judson Clark in 1905. Not unexpectedly, as the war intensified, Newman himself volunteered for wartime service and was replaced at St. Williams by J.H. White, Zavitz's former student and good friend. Before his departure overseas, however, Newman had been able to persuade Norfolk County to form a Reforestation Committee. The studies undertaken by the group would eventually guide the county's reforestation projects following the end of the war. It is interesting to note that Newman served in France with the Norfolk 133rd Battalion under the command of the member of the Ontario legislature who had secured support for the St. Williams Reforestation Station — Lieutenant-Colonel Arthur C. Pratt.

Before the interruption of the Great War, Zavitz had been dismayed at the slow rate of reforestation in Ontario, arising out of the general refusal of municipalities to adopt the County Reforestation Act of 1911. The province had also given municipalities the power to exempt farm forests from taxation in exchange for the owner meeting minimal forest-protection standards such as fencing off from livestock. This, too, had little response. Zavitz concluded that "public opinion [had] not been strong enough in any municipality" to exempt such forests from taxation.[4]

This gloomy outlook, on the eve of the First World War, was reinforced by ongoing difficulties in working with municipal councils. OAC President George Creelman, in an address to the Experimental Union, singled out these discouraging encounters after hearing a speech from Zavitz. Here, Zavitz expressed dismay at the fact that Southern Ontario's farming regions had less forest cover than the much more densely populated continental Europe.

Creelman told the OAC alumni:

> Now you men as citizens have got to take this matter up. Mr. Zavitz goes to a township council, or county council by invitation and it nearly always comes down to a matter of petty politics. It comes right down to the working basis that they are not willing to look five years ahead and sacrifice themselves for a principle such as we have heard here tonight. It is a municipal proposition, and you see, as members of the Experimental Union, you ought to be able to save this thing. I hope that every individual present will take this matter to heart, and as a citizen, will go before his township council, and persevere in his effort until the proposal becomes popular, and we get it well started. If you want to do something for your country as well as the

particular township in which you live, you cannot do better than to give particular attention to a matter such as has been discussed here tonight.[5]

While Creelman's words may not have sent a flood of farmers off to their township council meetings, they still had a significant impact. His call to action may well have been reverberating in the head of E.C. Drury when he accepted the leadership of the triumphant United Farmers of Ontario in 1919 and became the new premier of Ontario. Although Drury might not have been present when Creelman spoke that evening, as an active member of the Experimental Union at this time, he certainly would have read Creelman's remarks. As premier from 1919–23, he would use his position to do as Creelman suggested and encourage county councils, starting with his own in Simcoe County, to make use of these provincial programs. His efforts would be bolstered by the fact that many of the most effective members of his caucus and cabinet were OAC graduates, all with an understanding of the need for more forest cover in Southern Ontario.[6]

Having little possibility of changing attitudes in rural Ontario while in the midst of stopping a plague of forest fires in the north, Zavitz turned his attention as chief forester to agriculture lands on the Canadian Shield. These efforts were applauded by the chief inspector of the Board of Railway Commissioners, Clyde Leavitt (he also served as chief forester of the Commission of Conservation). Leavitt, originally from Michigan, had studied forestry at the University of Michigan, like Zavitz, and had worked closely with Gifford Pinchot in the establishment of the U.S. Forest Service in 1905 — the year Judson Clark left Ontario in disgust. Levitt was now spearheading the federal government's efforts to curb forest fires sparked by railways. He observed, "It is truly said by the Provincial Forester, Mr. E.J. Zavitz, that the only solution to this wasteland problem is the adoption of a policy, which shall have as its aim the gradual segregation of these areas to be permanently managed by some Government agency." This meant having forest lands "withdrawn from settlement" and having "established settlers ... transferred, so that "the lands should be permanently devoted to forestry purposes."[7]

The new premier of Ontario, William H. Hearst, who succeeded Sir James Whitney following his sudden death in office, had been born near Tara, Ontario (Bruce County). At that time, he was a Northern Ontario lawyer in Sault Ste. Marie. As a northerner, he held the belief that the Clay Belt of the north was among the best farmland in the world. While recognizing the area as having pockets of exceptional farmland for its latitude, Zavitz attempted to have him understand that even south of the French River "there [were] a number of large non-agricultural areas ... whose unfitness for agriculture has been realized." Such lands, he stressed, "should only be utilized for forest crops." He warned Hearst against "the widespread misunderstanding that the whole country is immediately fit for farming." Such delusions ignored "the depth of the overlaying mulch of peat layer" on top of the Canadian Shield.[8]

In 1914 Zavitz was able to persuade Hearst to have lands located around Algonquin Park and the Algoma Hills area, near the present community of Elliott Lake, "permanently devoted to forestry purposes." For the first time since the Conservative's 1905 election sweep, the government used provincial legislation to expand a forest reserve and added 4,080 square kilometres to the existing Algoma Hills Forest Reserve. The size of Algonquin Park was extended eastwards up to the limit of Renfrew County and included all the townships that had been partially added by Judson Clark's efforts in 1904.

Zavitz also persuaded Hearst's government to recognize this expansion of the Algoma Forest Reserve and Algonquin Park as an important step in protecting the last extensive areas where White Pine was still commercially viable and still successfully regenerating. Photographs of the newly expanded eastern half of Algonquin Park showed the White Pine was successfully re-generating along the Petawawa River. Although protection of the White Pine had been a goal of Algonquin Park's establishment in 1893, the heavy logging that had taken place in this newly acquired area prior to 1914 meant that this tree was no longer the dominant species. It was being replaced by a predominately hardwood forest and stripped of its former White Pine super-storey. As was the case in Southern Ontario, White Pine in the original western half of Algonquin Park had been reduced to unusual environments such as rocky lakeshores, islands, and exposed hilltops.[9]

The Highland Grove of White Pines in Algonquin Park, which later were cut down. Zavitz's work in the eastern extension of Algonquin in 1914, which ended the threat of the spread of agriculture there, helped to rescue the White Pine as a commercial species in Eastern Ontario.

From the time of Confederation right up to the First World War, White Pine had been the dominant pillar of Ontario's economy. During this period, payments for ancient old-growth trees harvested from Crown lands had been the largest source of provincial government revenues. By the eve of the First World War, Zavitz had become one of the world's leading authorities on White Pine, putting him in an excellent position to understand how the species had maintained a presence in the new eastern half of Algonquin Park, despite its having been logged over. He discovered the answer to this mystery was that this area had been logged before the penetration of railways into or near the future park's boundaries. With no trains for transportation, logging crews were forced to rely on tributaries of the Ottawa River, such as the Petawawa, which have their headwaters in Algonquin, to move the timber. Loggers could only afford to cut the most gigantic century-old pine trees and then float them down rivers to mills. As Zavitz explained to the government, "Many of the regions which were cut over in the earlier days of lumbering have produced splendid second growth White and Red Pine. In these operations, only the choice trees were taken and a large percentage of seed trees left. These areas testify to the ability of the Pines to hold their own in the struggle if given a reasonable chance."[10]

Zavitz correctly predicted that most of the White Pine stands in Ontario were doomed for two reasons. One was the eruption of fires caused primarily by agricultural intrusions into the Canadian Shield. The other was over-harvesting by loggers who did not leave behind sufficient seed trees for the pine to regenerate successfully. He warned that with "the present methods of cutting, when everything in the shape of a pine is taken out, it is not likely that we will see satisfactory new growth. Natural reforestation cannot bring back Pine in a region where no seed trees have been left. If Pine is desired upon these areas we shall eventually have to depend on artificial methods for forest planting."[11]

Part of the success of White Pine regeneration in eastern Algonquin was the exclusion of agriculture. Although the number of farmers here was quite small, their environmental impact was extensive. Like nineteenth-century pioneers, the burning of trees for ashes, which were used to manufacture soap and other industrial products, provided much of their income. Many of these farmers were squatters on Crown lands and were removed with the expansion of

A photograph taken by Zavitz of a White Pine grove at Lavrielle Lake in Algonquin Park.

the park. As Algonquin's superintendent, G.W. Bartlett, explained in his annual report for 1914–15:

> The squatters in the section recently added to the Park have been paid for their improvements and have all moved to other places. The section will fill up with game of all kinds, it being a splendid locality for such. There is a vast quantity of young pine coming on in many parts, making it doubly important to protect this section from bush fires which have already done considerable damage.[12]

Zavitz did achieve polices giving the White Pine in Algonquin his estimated "reasonable chance" for success. He persuaded the Grand Trunk Railway to employ a full-time fire ranger in the park. This "special man" was responsible for operating a tank car with a water pump and a "thousand feet of hose," and like a city fireman, he was always on duty. When a fire warning came, the fire ranger was "ready to be rushed to any point where right-of-way fires were to be found." Zavitz also had the railway rights of way in and around Algonquin cleaned up, having found that in the park and the surrounding Parry Sound District these were in a deplorable mess, which he described as "debris which had lain for years." Prodded by his authority, the railways quickly had them "cleaned up of all inflammable material."[13]

Before the Matheson Fire disaster, where there were over two hundred fatalities, only provincial parks and forest reserves had regular patrols by fire rangers to ensure observance of the fire prevention regulations. The extensions Zavitz secured for the Algoma Forest Reserve and Algonquin Park were now in a position to support the survival of White Pine as a commercial species. Now protected by fire-ranger patrols, these lands were no longer subjected to the massive forest fires that devastated northern farming regions such as the Clay Belt. Fires that burned soil down to the rock and destroyed its already limited fertility, however, were not effectively controlled until the mid-1920s, when the combination of stronger regulations, enforcement of same, and the use of aircraft in fire detection and suppression was developed.

Zavitz's success in saving Algonquin before this advancement of fire-fighting technology depended on the fire rangers' effective enforcement of his regulations. As the superintendent of Algonquin Park indicated in his 1914 annual report, commenting on its eastern extension along the watershed of the Petawawa River, there was a "vast quantity of young white pine" coming into maturity, but those were at risk from the "brush fires which have already done considerable damage."

In the summer of 1916, one of Zavitz's fire rangers responsible for patrolling the newly expanded area of Algonquin during its first fire season as a protected area was the now-renowned Canadian artist Tom Thomson. The painter, like most fire rangers, had been given a promotion after earlier meritorious service in fire-fighting work. Although this summer had the same heat and dryness that led to the horrific fires that devastated the Clay Belt, under Thomson's watch such horrors did not befall that portion of Algonquin Park.

It might be tempting to compare Edmund Zavitz and Tom Thomson, both with their love of nature, particularly trees, and both having relatives involved in the study of nature. In Thomson's case this was a cousin, William Brodie. He was an amateur entomologist and the province's first biologist for the newly established Provincial Museum (1903), which eventually became the Royal Ontario Museum.[14] Brodie had played a major role in convincing the Ontario government in 1893 to establish Algonquin as a provincial park. Until his death in 1906, he had been a driving force behind every conservationist effort of the provincial government. Some art historians claimed to see Brodie's mystical reverence for the natural world in the church-like settings of Algonquin's forest as captured in Thomson's paintings.[15] In reality, the fact that they both were involved in Algonquin Park at the same time is a coincidence, albeit an interesting one. Thomson was well-known to have supplemented his income by working as a guide and fire ranger in the Park. It is Thomson's mysterious death in 1917, and the ensuing intrigue surrounding it,[16] along with his revered position as an iconic Canadian artist that propelled him to fame, while Zavitz, by comparison has remained relatively unknown despite his major reforestation accomplishments and their impact on Ontario.

Zavitz's implementation of new fire regulations for Ontario included suppression of poaching in parks as well as increased fire protection. Thomson's duties as a ranger in the eastern Algonquin Park in the first year it was closed to agriculture, hunting, and trapping most likely would have made him unpopular with many, but whether this would be a motive for murder is very questionable. The fire regulations on logging companies, as Zavitz's reports indicated, would also have produced conflicts. Many did not even undertake such elemental precautions as the removal of slash from roads needed to transport fire-fighting crews.[17]

As his annual reports illustrate, Zavitz emphasized the need to maintain loyal and effective employees by better pay for rangers and an increase in status for the role. In describing what he needed in his rangers, Zavitz called for candidates being an "experienced woodsman," "energetic," and "able to deal with the public." Above all, he wanted the men to "be loyal to the organization of which he is a part." Such men, Zavitz stressed, "are not plentiful."[18]

With the help of dedicated rangers, Zavitz was able to perpetuate the dominance of the White Pine in eastern Algonquin Park. This success was documented by the University of Toronto forestry professor R.C. Hosie in his 1951 study on forest regeneration in Ontario. When Hosie examined Bishop, Freswick, and Anglin Townships in Algonquin Park, he found that pine regeneration under selective logging showed "better survival than in clear-cut areas." One reason cited for this success was that, in Algonquin, unlike the rest of Ontario, logging slash was cleaned up immediately instead of being allowed to decay over years. This cleaning up facilitated White Pine regeneration, which in other locations was delayed because the seedlings were buried in waste. Hosie also found that selective logging in Algonquin allowed the White Pine to become "established easily, possibly owing to the smaller opening in the stand, the smaller quantity of slash, and the presence of seed trees to selectively cut areas." Pine regeneration was best when cutting was done in a pine-seed year, and when, as Zavitz's reports predicted, a "minimum number of mature trees per acre" were left as seed trees. Algonquin's management also was helped by the Canadian Forest Service research station in Petawawa, some distance downstream from the park, which had developed a selective cutting system known as the uniform shelter-wood system. This also assisted pine regeneration.[19]

While saving the White Pine forests of the newly expanded Algonquin Park was no easy matter, the difficulties here paled in comparison to those that Zavitz faced in more northerly parts of the commercial forest of Ontario. The constant advice of J.H. White was of great benefit to Zavitz. Immediately after taking on the expanded responsibilities of the suppression of forest fires caused by railways, he appointed White as an assistant to administer the new fire regulations involving federally chartered railways that first fire season. Originally, he had been restricted to a staff of five in the Forest Protection Branch. Although the number of staff did rise, the big expansion would not come about until new legislation in 1917 permitted him to take over the rangers formerly employed by the Crown timber agents of the Woods and Forest Branch.

J.H. White and Edmund Zavitz, both of whom studied forestry as mature students, were close friends from the time of White's 1905 enrolment in Zavitz's class in Toronto until White's death in 1957. When White was away from his Toronto home, he would entrust Zavitz to check on it regularly, and, for this purpose, would leave a key underneath a flower pot near the entrance. The two men shared an enormous dedication to protecting and restoring forests and similarly shared a great deal of respect for each other.

During his student years in Toronto, White would do exhaustive fieldwork in the rugged forests of the Canadian Shield for the Commission of Conservation, the responsibility of Clifford Sifton, as minister of the interior. The resulting Trent Watershed Survey of 1913 documented the way fires were burning out the soil needed for the survival of future forests. As noted earlier, it was his dramatic photographs, exposing the rail-side burning of forests down to bare rock, that had been critical to the imposing of federal regulations on railways in 1912. An examination of the writings by J.H. White and Edmund Zavitz in government publications shows that their discussions would have likely been the best-informed commentary on the ecological problems facing Ontario during their over half-century of collaboration.

It was much easier to persuade railways in the tourist area of Algonquin and its environs to clean up their operations than in the rest of Ontario. With their tourism being heavily dependent on railway travel until the 1930s, a bare and blackened moonscape-like landscape, the result of repeated fires, was not a big attraction for Algonquin visitors. Zavitz had found, "Along many lines old logs, stumps, and other debris have been allowed to collect for many years. The condition frequently makes it difficult to put out small fires." To clean up the mess it was necessary to hire "special gangs" in "a number of districts" to undertake the work, in addition to the normal section crews.

Those railways, who had converted their steam engines to oil instead of coal, were exempted from the detailed regulations of their engines. However, since coal was still the norm for most steam railways, quite detailed inspections and regulations were imposed. A soft coal, known as lignite, was banned completely as a fuel for steam engines. For railways that used harder coals, a careful regime of regulation compliance and regular inspections were established. Coal ash-pans had to be wetted by overflow pipes attached to steam boilers. Steam stacks were required to be protected by a metal net to prevent sparks from flying out and igniting forest fires. Railways were compelled to keep a log book to record their adherence to the regulations. Zavitz's five inspectors read the log books every week and made sure that the condition of the steam engines met the requirements. In a way of diplomatically criticizing other railways, Zavitz praised the Grand Trunk for being the most co-operative in meeting fire-protection standards.

After a few years into the regulations, it became apparent that railways were exceeding the bare minimum his rules required. By 1915 Zavitz concluded that railways were "waking up" to the problem of forest fires, likely influenced by a combination of a need to reduce the damage that forest fires created for tourism, the higher costs of wood products, and the burning of potential timber that the railways could transport.[20] The success with railway regulations led Zavitz to warn against other forest-fire threats. He emphasized that, although logging slash was being removed from rights of way, it was still "in close proximity" to standing forests. He cautioned, "No matter how clean a right-of-way may be kept, it will be a physical impossibility to prevent forest fires when the fire hazards exist just outside the right of way." The other key challenge he identified was the threat of "settler's fires." In 1915 his inspectors found thirty-six "fires which were caused by settlers starting fires in a dangerous season and not controlling them."[21]

One of the benefits of Zavitz being a railway inspector was the right to travel on a free pass on federally-charted railways, anywhere in Ontario. This travel allowed him to witness firsthand the destructive impact of fire started by settlers. Since being prompted by Zavitz, railways had educated their employees about forest fires, and settler's fires were also being observed and recorded by locomotive engineers. Ernest Reed was one train engineer who went beyond his official duties of comprehensively documenting forest-fire dangers during the dangerously dry and hot summer of 1916. He saw farmers burn out forests on almost every trip over the first three weeks of that summer. When he attempted to reason with the farmers and discourage them from undertaking such dangerous burnings, they replied that they were fully aware that forest fires could get out of control and burn out homes. Such dangers, they believed, had to be accepted as the normal price for living in Northern Ontario.[22] In his 1915 annual report, Zavitz made a Cassandra-like warning of impending doom from forest fires. He noted that, during the year, the province had only been rescued from disaster from logging debris and settlers' fires "owing to the large amount of rainfall." Unfortunately, the summer of 1916 would be even hotter and drier.[23]

The term "settler" was a good descriptor to use in identifying fire threats to Northern Ontario forests during the First World War.

Zavitz's term is illustrative of the great difference in attitudes toward forests between First Nations people and the others living in Northern Ontario at that time. Native people were dependent on the caribou herds that required mature forests, rich in lichens, the same forests now being burned up. Protection of the caribou was one of the reasons that foresters in Quebec were determined to modify logging — a stark contrast to practices in Ontario. These "settlers," be they farmers, townsite developers, or prospectors, or mining corporations, all lacked such incentives to save forests from fire.

Interestingly, railways proved to be easier to regulate than the diverse "settler" interests. While federal regulations banned fire setting by railways right to the end of the month of October, and although he was given authority to regulate settlers' fires after 1917, Zavitz could never persuade the Ontario government to give him such a blanket prohibition that included October. It was in October of 1922 and 1937, respectively, that both the Haileybury and Dance (Fort Francis District) fires causing loss of life would take place.[24]

Farmers in Northern Ontario routinely used burning to clear lands for agriculture in the early twentieth century, bringing them into conflict with Edmund Zavitz and other foresters concerned about controlling forest fires. This is an image of a farm family in the vicinity of New Liskeard in the early 1900s.

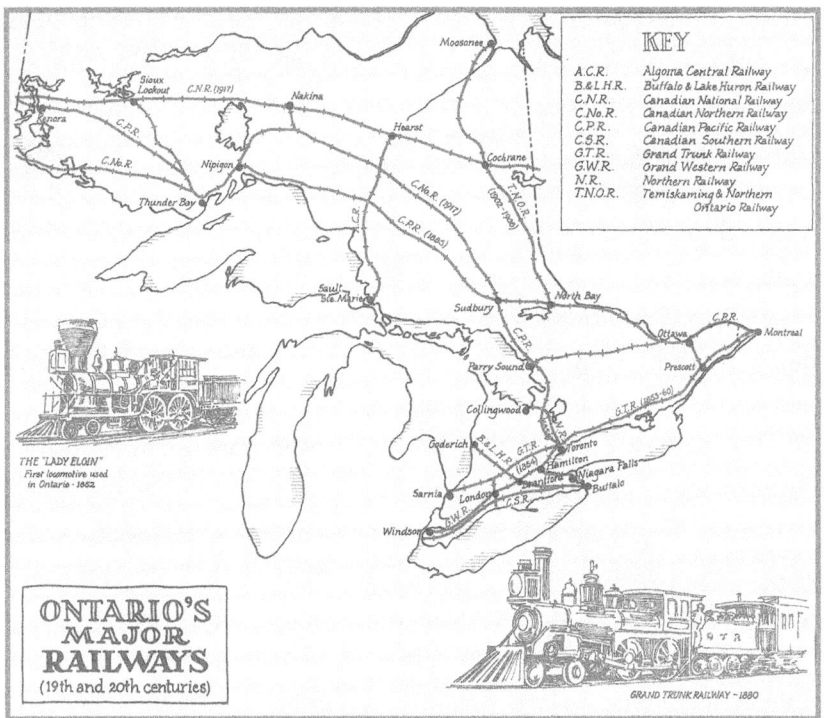

The spread of railway lines in Northern Ontario helped cause forest fires that burned soil down to the bare rock and devastated Woodland Caribou habitat. Zavitz's appointment as chief forester of Ontario was prompted by the determination of both Quebec and the federal government to curb the spread of such fires. Up to this point, Quebec was the only province to have a forest service run by trained foresters.

A combination of factors in 1916 made the summer in the Clay Belt the dreaded, perfect tinderbox setting for forest fires. During the excessively hot and dry season, high winds created a horrific conflagration out of numerous scattered settlers' fires. The winds were strong enough to cause church bells to ring without human assistance, as if in warning.

The Matheson Fire of July 1916 remains the worst in terms of fatalities in Canadian history, resulting in a total of 243 deaths. Over 2,548 square kilometres involving about twenty townships were destroyed by fire. The Town of Cochrane was burned out for the third time, resulting in fifteen deaths. Some eighty-one people perished in Monteith.

Twelve more died in open country. Eight perished in Iroquois Falls, and another nineteen were killed in Porquois Junction. The fire was named after Matheson, since this town, in addition to being totally wasted like the others, had the highest death toll. Approximately eighty-nine people died from fire in Matheson, although most of the doomed town's residents were able to flee to safety by train. The high death tolls were the result of the extreme difficulties facing fire victims. Flames poured into the town so quickly that they were described as being "like a howling tornado." After the first sparks landed, the town was engulfed in flames in only five minutes.[25]

The Matheson fire, for all its horror, in itself could not push Hearst's government to implement a stronger system of fire protection. Many factors would be required before the government would seek Zavitz's advice over new legislation. One was the drama over Tom Thomson's death in July 1917, with some press in the immediate aftermath reporting speculations that his suspicious death was caused by his efforts to uphold fire protection laws.[26] In effect, a number of obstacles remained. The fire regulations were so disliked by one part of the Ontario government — the Northern Development Branch of the Department of Lands, Forests and Mines — that it attempted to trivialize the implications of the Matheson tragedy. In response, this branch claimed that although the Rainy River forest fire of 1897 resulted in 140 deaths, it had brought about "one of the best agricultural sections in Northern Ontario."[27]

Zavitz, however, was helped by a series of articles in the Toronto *Globe* highlighting the weakness of regulations to prevent forest fires in Ontario. The reporting was topped off by a provocative editorial denouncing northern settlers as a dangerous law unto themselves, prone to reckless burning without regard to public safety. Public uproar in response to comments made by the minister in charge of forest-fire prevention, the minister of lands, forests and mines Howard Ferguson, gave the press exposé additional impact. He expressed surprise at the heavy loss of life, because in the fire zone there were many lakes and rivers where people could flee to safety.[28]

Ferguson turned to Zavitz to rescue him from a political crisis. In his memoirs, Zavitz recalls how, following the Matheson inferno, he was "called in by the Minister, to reorganize the Forest Fire Protection Service." At this time, as Zavitz recalled, "There was no permanent fire protection service. The problem was taken care of by two clerks of the Woods and Forest Branch. The details in the north were taken care of by the Crown Timber Agents whose staff went over to timber work in the autumn under the Woods and Forest Branch. The licensee or limit holders placed their rangers or agents as fire rangers, the Department having Inspectors who carried out supervision."[29]

Although Ferguson had Zavitz draft the Forest Fire Prevention Act of 1917, omission problems such as the absence of a blanket prohibition on October burning, as had been applied to railways, persisted. Zavitz was well aware of this issue; it is highly probable that the minister would have diluted the final terms to some extent before its introduction to the legislature. Despite such compromises, the law was a major change. Zavitz recalled:

> Rangers in all areas were appointed by the Provincial Forester; the licensee to pay one cent per acre [to pay for fire protection]. The system of Burning Permits was introduced. Power to order removal of fire hazards on private lands; power to regulate travel during dangerous periods; power to carry out the Act was placed under the Provincial Forester. Gradually, a permanent Forest Fire Service was established with districts administered by trained graduate foresters.[30]

There was another first. The Act also imposed regulations on provincially chartered railways that travelled through the Clay Belt — many had blamed them for the Matheson disaster.

What gave the Forest Fire Prevention Act its greatest strength was that, for the first time, Zavitz was able to regulate settlers' fires; the central feature was the permit system for such blazes. Between April 15 and September 15 of each year, such fires could not be set without a burning permit approved by a forest ranger. Zavitz was also given the power to remove fire hazards on any land, which in practice gave him the right, although at public expense, to clean up dangerous logging debris. He could also regulate travel during dangerous periods of time, which imposed restrictions on prospectors. His new powers extended to pollution control, since the chief forester was given authority over the disposal of mill and hazardous industrial waste.

As he had done earlier in establishing railway controls, Zavitz turned to J.H. White to implement the new forest protection service. White divided the northern portion of the province into fire prevention districts. After 1941 these became the administrative unit of the Department of Lands and Forests, reflecting the comprehensive approach taken for environmental protection by continental India and subsequently the U.S. Forest Service. The regional fire control districts, established on the basis of watershed and ecological characteristics,

This charming photo, taken by Zavitz in 1923, shows an unidentified young boy collecting pine cones for the Ontario Tree Seed Plant in Angus, Ontario, a program that still operates today.

formed the nucleus of the reformed department, which eventually would be controlled by foresters and other biological scientists. Each district had a chief ranger in charge of one or more deputy rangers. Under them were 1,400 fire rangers, who were supervised by inspectors, and the chief and deputy rangers.

A network of fire towers was established. Some of the tower men who staffed them won praise as local conservationists. One such tower man was Tom Parris, who served in Algonquin Park, and who would eventually be honoured by a memorial plaque testifying to his being a "natural born philosopher" who "protected the forests and its creatures." Zavitz supervised a new system of 3,218 kilometres of fire-protection roads and trails, with 827 kilometres being completely new. Portable fire-pumps and light pickup trucks were purchased, and long overdue moves made to obtain up-to-date technology.[31]

In his report to the 1917 meeting of the Commission of Conservation, Clyde Leavitt, the chief forester, praised Zavitz's work in expanding forest protection in Ontario. This he saw as the "new era in the matter of protecting Ontario's forests against forest fires." Leavitt was pleased:

> The Forest Fires Act has been remodelled on modern lines, and a forestry branch has been established, in charge of technically trained foresters, with full jurisdiction over the various lines of protection work. The total staff of the Forestry Branch at the height of the past fire season aggregated about one thousand men — easily the largest single fire protective agency on the continent.[32]

He also endorsed Ontario's decision to follow Quebec's example of a fire tax on those who had cutting licences on Crown land, in order to pay for fire protection. Leavitt saw that Zavitz had made an "excellent start." He went on to note with relief that a permit to regulate burnings by farmers was required, adding that "there have necessarily been some convictions for burning without a permit."³³

One of the compromises made in the tussle between Zavitz and Ferguson in formulating the new Act in 1917 was that the new legislation did not compel logging companies to clean up their slash. This problem was soon addressed in his annual reports. Here, Zavitz pointed out, "The fact that cut-over land and young growth make [up] of 54% of the total burned area indicates clearly the influence of slash and debris accompanying logging operations as a fire hazard. Forest protection can reach only a certain degree of efficiency without consideration of the matter of brush disposal." He also warned about loopholes regarding settlers' burnings. Although greatly reduced since the requirements of permits, there still were fires in areas not covered by the controls needed for forest fire suppression. Some ninety-one forest fires in 1918, he reported, took place "largely outside of the area where permits are required."³⁴

Despite the new powers and expanded fire-fighting service, Zavitz found that he was unable to lessen the massive scale of forest fires, although there was success in keeping them away from areas of human settlements. This bleak pattern emerges in his report for 1919–20, most of which describes conditions before the formation of the Drury government.

Aware of the role of fire in forest ecology, Zavitz stressed that natural fires caused by lightning were a tiny part of the forest-fire fiasco facing Northern Ontario. He found that only 3 percent of the fires of the summer of 1919 were caused by lightning. These, he stressed, "as a rule do not reach large proportions." Some 97 percent of fires were caused by various human sources, such as settlers' fires and summer logging. The fires were now so severe that "in quite a number of cases whole townships were swept over." In total, some 922,161 acres or 373,120 hectares in Ontario were burned out in 1919. Six percent of Ontario had been burnt.

To explain the scale of devastation, Zavitz wrote, "It is hard to realize what such an area is, but a conception might be formed by trying to visualize a strip of country six miles wide from Toronto to North Bay."³⁵ Zavitz also expressed concern about the breadth of fire devastation that hit valuable pine stands, destroying both mature timber and regenerating pine on cut-over lands. He found, "Included in the total burned over area are 237,226 acres of land classed as timbered, mostly with white pine. At the extremely low estimate of 1,000 feet per acre, this would mean as much timber as the Province received in dues last year. Heavy losses were also sustained by lumber concerns with the burning of camps, logging equipment and sawmills. These losses of course, must be ultimately passed on to the consumer."³⁶

Again, Zavitz stressed the problem of logging slash, which had eluded his efforts in drafting the 1917 Forest Fire Prevention Act. He found that the situation was worse than simply leaving the slash in the cut-over woods. Now, the roads so recently cleaned out or constructed at public expense to fight forest fires had become impenetrable because of their unauthorized role as dumping grounds for slash. He found a major problem in "the disposal of slash along main tote roads, about camps and dump grounds." This made it "necessary to pass over the roads which [were] frequently piled up with slash from the previous season's operations."³⁷ Such messes contributed heavily to fires that erupted from summer logging. These fires, although accounting for only 3 percent of the number of forest fires, were responsible for the large percentage of the province that was burnt up in infernos. Summer logging fires took a long time to put out, unlike the more quickly suppressed railway fires.

Zavitz also found that burning permits were being recklessly exploited by farmers and that penalties achieved in the twenty convictions were too low to deter offenders from such dangerous behaviour. He warned that the law should be changed to provide jail time for violations of burning permits, as had been successfully obtained in Quebec. He stressed, "A further amendment is required to cover deliberate defiance of the Permit Regulations. At present infringements can be punished by a fine only, and this becomes mere nominal, in fact cheap land clearing under certain conditions." He warned that in most years the dangers from settlers' fires in the Clay Belt were avoided only because "in an average year one counts on occasional showers which help to extinguish the dying fires which are seldom absolutely put out."[38]

With considerable determination and courage, Zavitz used his Forest Protection Branch to carry out functions that today are performed by the provincial Ministry of the Environment. In 1918 Zavitz launched an investigation into the role that air pollution from metal smelters was playing in deforestation around Sudbury. He arranged for the University of Toronto botanist Dr. J.H. Faull to investigate the impact "on the coniferous forest of sulphur fumes from winter roast beds located from four to six miles distant." These investigations pointed to an increased damage to trees when roasting was done "at temperatures above the freezing point." Zavitz found it strange that "there [appeared] to be no definite information in the literature based on experiments with respect to the susceptibility of white pine and others of our native conifers to sulphur fume injuries at the lower temperatures."[39]

The stress on Zavitz's duties during the First World War was increased by the eruption of the White Pine Blister Rust throughout North America — the result of the importation of infected nursery stock from Europe. Infestations of contaminated elm and chestnut trees would come later. Unlike these much more virulent devastations, however, the blister-rust problem was eventually controlled by research and co-operation between the continent's agricultural and forest services. As part of this effort, Zavitz attended a 1916 North American-wide conference in Pittsburgh addressing the problem and a later session the same year in Ottawa. Across the continent, a control strategy developed based upon the eradication of plants of the Ribes family — currants and gooseberries. Zavitz's staff at St. Williams eradicated these hosts around the Reforestation Station at St. Williams by hand. More widespread removal in the Canadian Shield, however, would have to wait until the 1940s, when herbicide applications were developed. Research eventually provided standards needed for Ontario to regulate the minimum distance between cultivated gooseberries and currants and forests.[40]

Zavitz singled out Walter Alexander McCobbin for praise in dealing with the blister crisis for his "very valuable assistance and co-operation." At the time, McCobbin was in charge of plant pathology at the Canadian Department of Agriculture's St. Catharines Research Station, which, at this time, was located at the former nursery of Delos Beadle, a pioneer advocate of reforestation from native trees.[41]

Following the drama of the massive forest fires in the summer of 1919, a positive turn in support of Zavitz's efforts to reforest Ontario came in October 1919, at the end of the fire season — the election victory of the United Farmers of Ontario (UFO). The win led to a sequence of events culminating in his close friend and fellow advocate of reforestation, E.C. Drury, becoming premier of Ontario.

Drury was not an isolated figure in support of policies of conservation. Within his governing caucus, composed of a coalition between the United Farmers and the Independent Labour Party, seven members were graduates of OAC. Of these, three — in addition to himself — were

Through working together to reclaim Ontario's land, Premier E.C. Drury and Edmund Zavitz achieved a remarkable recovery of much of Ontario's landscape.

in the cabinet: the minister of agriculture, Manning Doherty (Kent East); the minister of public works, Frank Campbell Briggs (Wentworth County); and a future premier of Ontario, Harry Nixon (Brant County). From their actions at the cabinet table, it was apparent they appreciated the lessons about the need for more forests in Southern Ontario during their Ontario Agricultural College studies.

The five unidentified men standing on a stump were posing for Edmund Zavitz just prior to the Oak Ridges Moraine being reforested. The area is now covered by the Vivian Forest.

— Six —

Drury and Zavitz: A Partnership

A MAN MOTIVATED BY deep Christian faith, E.C. Drury saw his task as the redemption of Ontario, making amends for abuses done to the natural world. When Drury described his friend Zavitz's appearance as an answer to prayer, he also expressed amazement at what they had achieved together. They had moved Ontario away from a rock-and-desert wasteland and back to productivity. Their accomplishments have been called "the coalescing of political power and professional capability in the persons of Premier Drury and Edmund Zavitz," which was "one of the rare cases in forestry when the combination of professional knowledge and political commitment established a process that provided long-term benefits to future generations."[1]

Drury understood the challenges he and Zavitz faced in attempting to change public opinion. As the premier later recalled, "At first public opinion was apathetic or hostile towards reforestation and many people probably looked at me as an impractical crank."[2] Drury and Zavitz had much in common. Both had their formal education, enhanced by what they learned in childhood from relatives, and had spent many hours in the outdoors. From his maternal relations, Zavitz learned about the destruction caused by deforestation of the Oak Ridges Moraine. Drury had heard similar stories about less-severe storms and flooding before the rampant deforestation. He had learned about farmers benefiting from summer-afternoon "four o' clock showers," which were replaced by torrential precipitation.

While taking a countryside walk near his farm, Drury was saddened to come upon a dried-up spring, which in his childhood had been "bubbling up" like "a boiling pot." As a child he had delighted in leaning down against a moss-covered tree to drink the pure spring water below. Now as premier, when looking around the once "merry little streamlet that gurgled away," he noticed that the formerly forested hillside that had nourished its waters was "stripped of trees." Drury saw much evidence of human abuse of the earth. He recalled, "One such storm in 1920, was a veritable cloudburst. If I remember rightly, it swept most of the bridges in the Township away and it cost more to repair the damage than would it have cost to have reforested the entire area in which the storm originated."[3]

Drury had ensured, before the 1919 election, that the reforestation of the deserts of Southern Ontario would be prominent in the United Farmers of Ontario's election platform. It was implemented, like most of its agenda. When in office, he pushed a number of conservation initiatives into being. The Department of Fish and Game was reformed to provide more protection from poaching. Harold Zavitz, son of Charles Zavitz, wrote a history of the Carolinian Lake Erie District for the Department of Lands and Forests. Here, he noted that before Drury came to power, such efforts were weakened through "the partisan spirit of political patronage." In his estimation, the UFO government produced "the first genuine effort to establish a protection service in wildlife conservation in Ontario."[4]

A problem facing the government, however, was how to foster long-term political activism by its members so that the intense involvement of farmers in politics would not be a one-time fluke. A massive party picnic was organized and held in Rondeau Provincial Park, beneath the shelter of the towering old-growth trees so often photographed by Zavitz. The gathering attracted 12,000 supporters, breaking its all-time attendance records. Historical accounts indicate that the involvement of farmers in the UFO party went up, and their member for Kent East, Manning Doherty, the party's capable minister of agriculture, was re-elected.

In 1919 Zavitz was soon able to bring his co-operative planting program with farmers up to the pre-war standard of 300,000 trees planted annually. He understood that the "greatest demand from farmers for planting material" was from the St. Williams Station area, "proving that actual demonstration is the best form of education."[5] Following Drury's election, Zavitz was able to purchase another 106 hectares to expand the forest around the Station, bringing its size up to 729 hectares. Its superintendent, Frank Newman, who had returned to his former post after the war, immediately embarked on a number of efforts to attract visitors to St. Williams. He operated the entrance to the station as a scenic park, which included picnic facilities and ball fields. He also used speaking engagements, newspapers, letters to individuals, and, of course, Zavitz's favourite technique, demonstration tours designed to win converts to reforestation.[6]

By the time of Drury's election, Zavitz, with the help of local friends such as Monroe Landon, had won the support of local councillors, and municipally administered reforestation projects were underway in Norfolk County. Although no longer in the legislature, his friend Arthur Pratt was now in constant touch with these councillors through his new role as sheriff of Norfolk County. As Drury would later do in Simcoe County's courthouse in Barrie, Pratt would use available time in the Norfolk Courthouse in Simcoe to encourage councillors to purchase land to protect and expand forests. Their first purchase, under the 1911 Counties Reforestation Act, was in 1922. By 1947, Norfolk County had acquired 746 hectares and had moved from halting the advancing sand deserts to dealing with the erosion of stream valleys that had been stripped of forest cover.[7]

Although attitudes toward conservation were beginning to change in Norfolk, both Zavitz and Drury were aware of the unaltered reluctance of county councillors in other parts of Ontario to use the 1911 Counties Reforestation Act. Obviously, more reforestation stations were needed in Southern Ontario, to show skeptics reforestation's power to transform barren deserts. Drury later recalled:

> Shortly after I took office, E.J. Zavitz, the Provincial Forester, and I got together on the subject in which we were so interested in: reforestry in older Ontario.

After the establishment of, many years before, the first Provincial Forest Station at St. Williams in Norfolk County, the Government had lost interest in the matter and very little progress had been made. Zavitz and I laid plans to invigorate the project.[8]

To invigorate reforestation, the Ontario Seed Tree Plant was established in Angus, Ontario, in Simcoe County. It still exists today. At Angus, seeds were extracted from cones and planted in nursery conditions to become seedlings for sale or shipping. Homegrown stock was produced to avoid further disasters stemming from importation from the United States, namely the issue of White Pine seedlings grown in American nurseries from stock contaminated with blister rust, which had come from Germany. Much of the early work of the station was to encourage selective breeding from trees that had shown resistance to this disease. The Ontario Seed Tree Plant also sought to conserve genetic material from the various climate zones of Ontario, since seedlings from seeds collected in their distinctive regions grow faster and survive longer in their native climate zones. Zavitz and Drury understood that the difficult task of extracting pine seed from cones required specialized equipment and a network of suppliers, such as farmers harvesting cones from their own forests. Although St. Williams had a seed tree plant, which would continue to operate into the 1950s, it was inadequate for the massive scale of the reforestation program that was being planned.

The seed tree plant was established in 1922 on a desolate sand wasteland near Angus. Like the tree nurseries elsewhere, it developed into an impressive demonstration forest of some 777 hectares and eliminated the sand waste. Its buildings, surplus aircraft hangers from the end of the First World War, had been transferred from Camp Borden. Initially, the plant had the capacity to handle twenty bushels of Red Pine and seventy-five bushels of White Spruce per eight-hour day. Two men at the plant did all of the work, including, as Zavitz noted in his 1922 report, "firing, turning and filling the drums and cleaning the seed." In its first year, the plant gathered 444 bushels of Red Pine, 175 of Jack

Pine (obtained from a northern logging operation), thirteen bushels of Red Oak acorns, and smaller amount of Black Locust, Black Cherry, Sweet Chestnut, Black Walnut, and soft maple seeds.

Zavitz issued a circular entitled "Gathering Red Pine Cones for the Ontario Government" and distributed it freely throughout the province. Supported by newspaper ads, the publicity eventually "reached hundreds of people interested in this work." While severe weather conditions produced relatively few cones at first, in 1925 "a bumper crop of 3,126 bushels of pine cones" was collected.[9]

The White Pine Blister Rust erupted during the First World War, but it took some time before the seriousness of the situation was understood. In order to best continue the reforestation projects already planned, however, Zavitz selected Red Pine as the most appropriate, since it could survive in the hot and dry conditions of the arid sand wastelands. He found that "our native pine ... grows well on poor soil, has few enemies, and is valuable for timber purposes." He viewed the lands around Angus as the perfect place for collecting Red Pine cones, believing that it was "doubtful whether a more suitable area than this for collecting red pine seeds exists in the province."[10]

Following the establishment of the tree seed tree plant at Angus, Drury and Zavitz met there on a weekly basis for about a year and a half to plan the next stages of reforestation in Ontario. This arrangement worked well, as on Friday nights Drury would make a detour to nearby Angus on the drive to his farm north of Barrie, and Zavitz, who was often out on field work, would plan his schedules accordingly. High on their priority list was the establishment of more reforestation stations. The one they had already selected in Midhurst in 1905 was now operating under the supervision of a recent graduate of University of Toronto's forestry school, Isaac Marritt. The other station was to take on Zavitz's long-term mission of reforesting the Oak Ridges Moraine. Located on the moraine at Orono, its operation was entrusted to another graduate of this school, George Linton.[11]

Two transplant nurseries were also established. One, in Prince Edward County on the site of the future Sandbanks Provincial Park on Lake Ontario, was there to stabilize the shifting sand dunes, which were the result of extensive deforestation. The other was located in Kemptville to encourage reforestation throughout southeastern Ontario. With its close proximity to the neighbouring Kemptville Agricultural College (the campus was created as result of Drury's government initiative, under leadership of his minister of agriculture, Manning Doherty), the nursery became part of a reforestation education program. The college established an eight-hectare mixed forest along the Ottawa–Prescott highway.[12] Drury's government also established more provincial parks: Long Point, Severn River, and the later abolished Franklin Island Park in Georgian Bay. The latter park, before its closure by the Hepburn government, was used as a summer camp by the Federation of Ontario Naturalists.

Drury and Zavitz, working closely together to have Algonquin Park become a model of good forest management, secured regulations that established controls over logging slash. Another of Drury's reforms was a ban on beaver trapping in the park, which was a move that played a critical role in restoring the species in the province. Still, more reform was required. In a 1921 report, Zavitz warned, "Until brush disposal is systematically undertaken as an integral part of the operations of logging, our forests will burn." An amendment to the Provincial Parks Act of 1921 added the requirement for compulsory brush disposal as a condition in timber licences.

An experiment was conducted to determine the cost of such measures. Two gangs of men, under the supervision of the deputy chief fire ranger, disposed of the collected slash through a controlled burn. As a result, Zavitz determined that, in pine operations, it cost from 70 cents to $1.00 per thousand feet of pine to clean up slash. Pulpwood operations with usually smaller trees could be cleaned up at the cost of 40 to 50 cents per cord of wood harvested.[13] Budgets for slash removal could now be set. As a bonus, slash disposal helped in the regeneration of the important stands of White Pine in the eastern half of the park.[14]

A 1921 survey employing sixteen University of Toronto forestry students enabled Zavitz to document the importance of Algonquin Park as the last stand for the White Pine in Southern Ontario. The survey, known as the "Ottawa-Huron Survey," documented how, in a region where Red and White Pine were formerly the predominate species, now only 3.7 percent — some 123,000 hectares— of the area still had coniferous forests. Zavitz discovered that "Almost one-half of this, is again, located in Algonquin Park." The park comprised three-fifths of the forest still suitable for commercial logging south of the French River and amounted to only 26.4 percent of the forest area. Outside the park, some 10.6 percent of the land was also threatened by "a particularly vicious fire hazard to commercial stands owing to the accumulated slash following recent logging operations." One-fifth of the Parry Sound area forest was blighted by such slash.

In addition, much of the substantial mixture of pine and other species was heavily concentrated in Algonquin's forests, and some 41.6 percent of the second-growth poplar-birch forests that contained a substantial pine admixture were in Algonquin. The survey also found the Madawaska River watershed, downstream from the park, was "in a semi-barren condition, supporting, in the main, poplar-birch degenerates with an occasional fire-scarred pine." Muskoka, which Zavitz found was marked by "the presence of so many abandoned farms" had "practically no intermediate or second" pine growth. Bleak barrens were the norm in parts of Renfrew County immediately "adjoining the timbered region of Algonquin Park."[15]

In 1921 Zavitz and Drury developed a new element in their reforestation strategy — the demonstration forest. It was seen as a simple way to get municipalities involved in reforestation. All that municipal governments, school boards, or other public institutions had to do was to set aside a small amount of land and fence it off. The Department of Lands and Forests, under Zavitz's supervision, would then reforest the property. The program nicely complemented another popular initiative of the Drury government, the creation of Municipal Parks Commissions. Municipal reformers, complaining about corrupt municipal councils not spending enough money on parks, had been lobbying the provincial government for this for some time. Once established, these parks commissions would be required to spend a certain percentage of their assessment revenues for park purposes.

John Squair, always most interested in his nephew's career, was able to document the impact of demonstration forests on the Oak Ridges Moraine. One such forest was established at the corner of Scugog Road and the Fourth Line Concession of Darlington Township. After the township purchased two hectares for $300 "in the following spring," the land was "planted with young pines obtained from the Ontario government nurseries." Squair was delighted how the Orono Reforestation Station had become "a source of supply to cover many a bare, unsightly spot with beautiful and useful wood." He noted the names of the first farmers to obtain trees from the Orno Station: John Stewart near Kendall, and John Tamblyn and A.J. Staples near Leaskard.[16]

Zavitz saw the demonstration forests and the forests around government nurseries serving an important task of public education. The nurseries would "not only reclaim waste lands but [would] supply planting material to private owners and [would] be a local demonstration of reforestation" as had been achieved around St Williams. The forests there "have become splendid demonstrations of what may be expected from work of this kind."[17] In Eastern Ontario, a demonstration forest began with a one-hectare park in Mountain Village in Dundas County. In Leeds, in Elizabethtown Township, a "sand ridge that had been inconveniencing traffic for many years" was reforested as a demonstration to neighbouring property owners "threatened with drifting soil."[18]

One of the cities where the demonstration forest was used most effectively was Hamilton, a bastion of labour support for Drury's government. The chair of Hamilton's Park Commission was the dynamic Thomas McQuesten, who used the program effectively to expand forested parkland in the city. A similar use was made by the St. Catharines Parks Commission. In Uxbridge, the creation of parks was spearheaded by Owen Davies, the deputy reeve, who was so inspired by reading a Drury speech on reforestation that, on the following day, he recommended that the Uxbridge Council "purchase land for reforestation purposes."[19]

— *Drury and Zavitz: A Partnership* —

The Demonstration Forest Program took root in Simcoe County, one participant being the Canadian military at its Camp Borden base. Barrie joined in along with neighbouring Innisfill Township, where a twelve-hectare park on the shores of Lake Simcoe and a planting on Yonge Street were carried out. Sunnydale Township created a four-hectare forest near the community of New Lowell, where Zavitz had found that the "drifting sand from the adjoining fields ... practically blocked" one section of the Glen Cairn Road.[20]

By reforesting the edges of eroded streams and hill slopes as a model, Drury used Beeton's successful Demonstration Forest Program to restore its threatened water supply. Earlier deforestation had caused Beeton's springs to dry up, but now with more forest planted, the water flow exceeded what village residents could remember. Drury hammered the point home, stating, "In our own County, the Village of Beeton has increased its water supply by a bit of judicious reforestation, and at Midhurst Park, the springs have increased their flow for the same reason." Midhurst Park, another Simcoe County project, had also been successful in protecting the source of their drinking water.[21]

Despite all evidence of the benefits of reforestation, Drury was still finding many skeptics among the political elite of Simcoe County. His discovery of the depths of this hostility marred the otherwise joyous occasion of the opening of the Midhurst Reforestation Station in 1920. Here, Drury received a dressing down by the county warden, who lectured him and said, "What's all this nonsense about spending public money on growing trees ... if you want a few trees, why don't you go into the bush and get them." The rebuke was especially significant, since the warden was the editor of a weekly newspaper and had been principal of Simcoe County Collegiate Institute.[22]

The warden's complaints had underscored the need to develop new incentives in the Agreement Forest Program. While municipalities still had to purchase the land for reforestation, the province assumed the cost of planting, maintaining the trees, and fire protection, and large forests would have a paid custodian employed by the Department of Lands and Forests. The province could also pay the municipalities back by partly reimbursing them through assuming joint ownership. In addition, municipalities could borrow money from the province to purchase the land and not be charged interest.[23]

To assist in countering the prevailing negativity, Drury initiated a publicity campaign and Zavitz hired Arthur Herbert Richardson, a recent forestry graduate from Harvard, as its manager. Like Zavitz, he had studied biology at McMaster University as an undergraduate and possessed a deep understanding of Ontario history. Richardson mounted an effective newspaper advertising campaign, stressing both the economic and ecological benefits of reforestation.[24] He was helped by Midhurst's beauty (the reforestation station having produced 1,820,000 trees in its first year), even though its pine trees had been planted in what Zavitz described as "forty-five acres of old stump land."[25]

In time, Drury was able to wear down the resistance of Simcoe County's councillors to reforestation. During his time as premier, he saw the county approve the purchase of "1,000 acres, the legal minimum" for the first agreement forest in Ontario's history, the Hendrie Tract, located thirteen kilometres from the premier's farm at Crown Hill. The first tree was planted on May 8, 1922. Over the next two years, some 1,244,600

A 1923 photo of the Hendrie Forest site, the first agreement forest in Simcoe County, near Midhurst, was taken just before planting began.

trees would be planted there. In the first year of the plantings, a plague of grasshoppers "played havoc" with the young seedlings, destroying many of them. However, after the surviving trees began to change the site's desert-like conditions, there were fewer problems. Simcoe County expanded the forest and today it covers some 1,214 hectares.[26]

During the first seven years of its participation in the Agreement Forest Program, Simcoe County planted 230,900 trees annually. It also used the earlier county forest legislation to purchase smaller tracts without the help of the province, acquiring 161 hectares by 1949. In the 1940s, planting had reached 454,577 trees per year. A few years after the first Hendrie Tract was reforested, a second agreement forest was purchased at Orr Lake and eventually came to 404 hectares. In 1948 the first thinnings to increase bio-diversity and establish more natural conditions were carried out at both the Orr Lake Forest and the Hendrie Forest. For the first time, after waiting for twenty years, Simcoe County began to receive some revenue from logging. Simcoe County's forest network now involves 150 tracts comprising 12,545 hectares of forests.[27]

Even after a century of clearing the land, giant pine stumps remind us of the spectacular forests that once covered Southern Ontario. This image, taken in 1980 with the photographer's son Brandon Borczon perched on the stump, was taken on the Oak Ridges Moraine near the Headquarters Tract, the former Vivian Forest in York Region. It is now part of today's York Region Forest.

Apart from Simcoe County, the only other county in Ontario to show interest in the Agreement Forest Program while Drury was premier was York. In 1920, Zavitz had met with the York County Council to discuss reforestation. A receptive audience, the County Council had been concerned about the issue of blow-sand conditions on the Oak Ridges Moraine since 1910. Waterways were drying up and heavy rainfalls were causing sudden, destructive floods. Game and fish had vanished. Two years after this meeting, the county purchased roughly thirty-two hectares in Whitchurch Township. This tract was later incorporated into the second agreement forest in Ontario, the Vivian Tract.[28]

The Sandbanks Transfer Station in Prince Edward County, developed during Drury's government, became the focus of one of the most challenging tasks of reforestation in Ontario. Zavitz described the sorry state of the site: "Much excellent farm land has for years been continually spoiled or covered up by drifting sand." He estimated that "eighty-five acres of lands once well forested with White Pine and Cedar were now covered with drifting sand." Additionally, the area was challenging to reforest: "The planting material used in this work consisted of Carolinian poplar cuttings and limb material. Belts were planted one hundred feet apart, each belt consisting of one row of limb material and five to seven rows of cuttings planted at right angles to the prevailing winds." The planting covered the banks "where the despoiling of agricultural land" from drifting sand was found to be "most imminent."[29]

Zavitz was careful not to reforest any of the natural sand dunes in what later became Sandbanks Provincial Park and left some human-created ones alone, perhaps to provide examples of abuse to the earth. A gruesome cedar skeleton in the sand provides an iconic photograph for the park, whose restoration plan notes, "Some large cedar skeletons in the northwest remain as evidence" of past deforestation. At the site of today's Sandbanks Park, Zavitz had eighty hectares reforested, an amount sufficient to stop its shifting sands from burying roads and orchards. A combination of timber cutting and cattle grazing had destabilized the soil, causing sand dunes that were originally confined to the coast to migrate inland. In 1881 the West Point Road was buried under 30 metres of sand; the West Lake Brick Company's factory, in the community of Athol, was similarly assaulted and had to be moved.

As noted, the restoration project proved to be most difficult. Zavitz described how "only trees that will grow under the most adverse conditions are of use in work of this kind … [to] stem the march of the sand." The biggest challenge was "to prevent the sand from covering the planted area during a severe winter storm," when the sand mingled with snow drifts. In the summer of 1922, Zavitz had 609 metres of plank fence and 914 metres of lath and brush fence erected as catch fences. These, he reported, "were erected at regular intervals and at right angles to prevent the greatest drift … to prevent excessive drifting and give the trees a chance to become established." Sweet clover, obtained from local farms, was spread over the sand for stabilization. Over a fifty-year period, three million trees were planted with extensive mulching used to trap any shifting sand. Later stages of reforestation included Hackberry trees in the mix to support butterflies dependent on this native species.

The major difficulty of the half-century victory over the sands is illustrated by the ongoing fragility of the site. The provincial park management plan notes how "the entire site is ecologically fragile: the substrate is almost pure sand, with a thin layer of organic material that is mainly composed of pine needles." It also prevents the use of any controlled burns, since any fire outbreak now will "destroy the very thin layer of soil and duff (decayed organic material) present and create open conditions too quickly to allow for naturalization." In 1960 the forest Zavitz restored was incorporated into the 1,600-hectare Sandbanks Provincial Park. This forest is gradually becoming a more diverse mix of native species, such as White Pine, Chokecherry, and Eastern White Cedar. Cedar seedlings and saplings are planted in open areas and around the forest's edges to speed this process, and the invasive exotic Common Buckthorn is controlled through mulching.[30]

Many foresters that Zavitz encountered at international forestry conferences told tales of spreading sands similar to those he witnessed on the wind-swept shores of Lake Ontario. The most lengthy of these conferences, the Imperial Forestry Conference of 1928, fostered a deep bond among foresters in the British Empire. It found Zavitz on a cruise throughout the South Pacific, in the company of his wife, Jessie, and in the presence of the dedicated director of the Canadian Forest Service, Ernest Finlayson. In New Zealand, he was astonished to learn that the type of policies for reforestation he and Drury had recently developed had been implemented in 1883.

Zavitz envied other imperial foresters with reforestation, not fire suppression, as their most important task. He told one gathering that they must feel it strange that the "prominence of forest fire prevention" in Ontario was "so very noticeable" and that he longed to share in experiences where "the peaceful fields of forest investigation or reforestation" were more prominent.[31]

While in New Zealand, he was impressed by the size of the Maori reservation land, some 25,495 hectares in extent and "nearly all forested." Many of the imperial foresters were empathetic towards indigenous people, as demonstrated by British forester Richard St. Barbe Baker, who, while in Kenya, bravely interposed himself between a local African man and a British official who was assaulting him.[32]

Baker founded a visionary organization, Men of the Trees, in Kenya in 1921. Now under the leadership of Prince Charles, it is known as the International Tree Foundation, with headquarters in West Sussex, England. Although British, Baker had extensive ties to Canada. A graduate of the University of Saskatchewan, he became a lifetime friend of his schoolmate, John George Diefenbaker. When Diefenbaker became prime minister, he did encourage reforestation, especially around Ottawa. From Baker's perspective, reforestation was the moral substitute for war. Through his private charity Men of the Trees, which was supported by powerful and wealthy British aristocratic conservationists, he began making plans in the mid-1920s for armies of over two million men to be transformed into a reforestation corps for the Shahara. He had envisioned British-French efforts on the "green-front," but the Second World War would intervene. Baker, however, did have some major successes. His Men of the Trees effort had a major impact in the part of Kenya where he worked, the only part of Kenya that was not ravaged by the Mau Mau rebellion in the early 1950s. Men of the Trees would later come to Ontario and, by mobilizing World War One army veterans, provide critical support for Zavitz's push for conservation authorities.

This non-violent yet militarized effort is reminiscent of Zavitz's approach to building morale in his Forest Protection Branch — what by then was the Department of Lands and Forests (Drury had created a separate Department of Mines). Zavitz poured considerable passion into engaging his employees with their mission. He would gather together foresters involved in reforestation and strategize in a war-room-like setting so, like Baker, they could go forth and stop the advance of the desert. The resulting disciplined unity of purpose, however, did not extend north of the waters of the French River. Drury's government, during its three-and-half years in power, failed to strengthen Ontario's forest-fire prevention regulations in the northern reaches of the province, despite Zavitz's well-reasoned pleas in his annual reports. Zavitz expressed increasingly more concern around the issues of slash removal, enforcement of regulations, and heavier penalties for illegal forest burning. In one report, he informed Drury's government that "much of the barren land ... some 439,383 acres that had been burned that year" was "once covered by merchantable timber." It had been "reduced to rock desert upon which merchantable timber will not again be available for very long periods of time." This devastation, he warned, was the consequence of "repeated burns."[34]

Drury set up a Timber Commission to investigate his suspicions of corruption among northern logging interests. Subsequent convictions, obtained through the courts, provided substantial revenues for the provincial treasury. The Commission exposed fraudulent use of mining claims to seize timber on Crown lands without payment to the government.[35] It is not surprising, then, that both the government of the time and the forestry regulations it was trying to implement became even more unpopular. What galvanized the northern opposition to scientific forestry was outrage against "southern" comments regarding the suitability of the Clay Belt for agriculture. These comments were made by the University of Toronto's dean of forestry, Bernard Fernow, and published in a report issued by the federal Commission of Conservation.

Fernow set off the political storm in the north by observing that the Clay Belt was best suited for pasture, not appropriate for most field crops, and that forested tracts should be reserved within it. His leading critic, Robert Roswell Gamey, a Conservative MPP for the Riding of Manitoulin, had served in the Ontario legislature from 1902 to 1917. An insurance agent and mining speculator, Gamey was a colourful character. Through his mastery of political intrigue, he became the proverbial thorn in the flesh of Fernow and professional foresters in general. In response to Fernow's Clay Belt report, Gamey induced all northern legislators to denounce Fernow as "the favoured child of the Whitney government" — a surprise to Fernow, as whenever he sought to advise Minister Frank Cochrane, he was rudely shown the door.[36]

Gamey's attacks on foresters were popular in his northern riding of Manitoulin, where agriculture was on the verge of collapse, its fragile soil being an extension of the Niagara Escarpment. At that time, the land was being cultivated largely through the use of horses and was marginally economical. This farming constituency applauded Gamey's attacks on foresters and kept him in the legislature from 1902 until his death, in 1917, despite many parliamentary censures. His sudden death provided the basis for the first United Farmers of Ontario representation from Manitoulin in the provincial legislature, through the election of a farmer and former Mennonite missionary Ben Bowman. When the UFO formed a government in coalition with Labour, Bowman was the party's most experienced legislator and its only representative from Northern Ontario. This background led to his becoming minister of lands and forests. Henry Mills, representing Fort William, was one of the two northern members of the Independent Labour Party, and he became minister of mines.

Unlike many southern UFO members, Bowman lacked the Ontario Agricultural College exposure to the principles of conservationist forest management. He was disliked by one of Drury's closest friends, the farmer-professor C.P. Sisson, on whose farm Zavitz had done reforestation work in 1906. In an article published in *The Canadian Forum*, where Sisson deplored Bowman's "lack of imagination," he was likely re-creating conversations between himself and Drury held in private on their farms.[37]

Drury faced many calls to turn the management of northern Crown lands over to professional foresters instead of the politically influenced Crown timber agents in the Woods and Forest Branch (renamed Timber Management Branch) of the Department of Lands and Forests. In particular, Robin Black — the secretary of the Canadian Forestry Association — was pushing for this reform, even though logging interests on his own board were opposed. Black had the support of the deputy minister of lands and forests, Albert Grigg, who earlier had helped Zavitz secure adequate office space at Queen's Park. With Grigg's support, Zavitz was placed in charge of Timber Administration, a role that would only last for two weeks.

Bowman, who with his northern farming background saw the burning of forests as a legitimate way of clearing off trees for farmland, was opposed to Zavitz taking charge of the Timber Administration Branch that actually managed the cutting rights to trees on Crown lands. Unfortunately, after Grigg's retirement just two weeks later, Bowman appointed Walter Cain as the replacement deputy minister of the department; he was a former school principal from Newmarket without university-level training in the natural sciences or forestry. Zavitz became fed up from all of the futile "fighting" with them, and he told Drury that the control of the Timber Management Branch was not for him. He added, frankly, "I never got anywhere." When Drury responded, "You are not standing up for your rights," Zavitz said, "Look, Mr. Drury, do you expect me to stand up against your Minister and Deputy Minister?"[38]

In response to Zavitz's complaints, Drury hired Judson Clark — who had been in British Columbia all this time working for private companies in forestry — as an independent consultant to conduct an inquiry into the Department of Lands and Forests. As part of his findings, Clark urged that the Lands Branch, heavily involved in the promotion of agriculture in Northern Ontario, be put into a separate department. Lands that were to remain Crown lands would be administered by a new Forest Department under the direction of the chief forester of Ontario, Edmund Zavitz. The new department would be in charge of Timber Administration, Forest Protection, provincial parks, and be merged with the Fish and Game Department.

Zavitz worked in close co-operation with Clark during the preparation of the report. Clark did point out, "I am glad to pass on the suggestion of the provincial forester" to prevent fraud in public revenue for timber cutting. The recommendation was to ensure that "all shanty books," where the Crown timber agents recorded the volume of timber cut on Crown land, "be serially numbered and made readily accounted for at the end of the season."[39] To persuade Drury to act quickly and put Zavitz and his team of professional foresters in charge, Clark pointed out the earlier success they had in curbing forest fires from railways. Using data from Zavitz's annual reports, he highlighted how railway fires had dropped from causing 49.5 percent of forest fires in Ontario in 1917 to only 14.8 percent in 1921. He also documented the impressive impact of the 1917 Forest Fires Prevention Act in cutting railway fires after the provincially chartered lines were brought under Zavitz's inspections. When this reform began in 1917, some 28.5 percent of locomotives were found to be in non-compliance with regulations. By 1921 that number had dropped to 8.3 percent.[40]

That Clark was not able to spur Drury on to quick action (although much of it, such as the creation of a Department of Forests in 1926, was implemented later), is illustrative of how forest-fire disaster was too frequently not high on the government's agenda. This time the failure to bring about the recommended reforms would see the disastrous Haileybury fire of October 4, 1922, which caused forty deaths and the destruction of 6,000 homes.

During Drury's government, Zavitz had warned the province that despite the improvements of the 1917 legislation, they had only narrowly avoided another town being destroyed by fire in 1921. Interestingly, just prior to the Haileybury fire, in a tribute to his fire rangers, Zavitz had explained:

That no towns or settlements were burned and no lives lost was due in some instances only to the morale of the field staff and their work with pumps. During early June five northern towns were seriously endangered, but in each case it was possible, with the use of two or three pumps, to [control] the fire before buildings were destroyed. In one instance the situation became so critical that the women and children were placed on a special train ready to leave the town, but the rangers, with their faces muffled with wet clothes so hot was the blaze, were able with three pumps to check the fire and not one building was burned down within the town limits.[41]

In addition to all his concern about the reforestation needs of Southern Ontario, Zavitz consistently stressed that the fire danger in the north was a critical emergency. In his 1920–21 annual report, Zavitz warned, "The outstanding feature of forest administration in the Province, as in all Eastern Canada, is an inability to control the losses from forest fires. The undertaking is so large and its bearing so important that the other phases of administrative work are comparatively minor matters."

Prior to Drury's election in 1919, a touchy political issue was discovered in the elaborate forest-fire prevention regulations J.H. White had developed for Zavitz. The blueprint had taken him thirty months to organize while on leave from the forestry faculty of the University of Toronto. The Little Clay Belt around Haileybury was originally included in this system of forest-fire prevention districts based on watersheds — a feature regarded as an insult by farm, business, and municipal leaders in the region. The mastermind behind the stirring of the political pot was the Conservative member for the riding of Timiskaming: Thomas Magladery. In his memoirs, Zavitz recalled him as being supported by many northern "township leaders." As a result, the Little Clay Belt was removed from the fire-prevention regulation.

The traditional burning methods to clear forests were targeted on the remaining pockets of reforested wetlands in the Little Clay Belt that had so far escaped the plough. The removal of these controls meant that the burning by farmers in this district in October 1922, which erupted in disaster, was legal.[42]

Forty years after the catastrophe, Zavitz recalled:

> The farm land contained considerable muskegs [boreal forest swamp wetlands], and owing to a very dry September considerable clearing of these was carried out. The conditions were alarming by September and the Provincial Forester instructed the Cochrane office under Mr. Poole to visit the area and find out if the Rangers should again be placed in the area. This was objected to by the leaders of the district. On October 4th the area was swept by a terrible wind. A muskeg west of the Station had been burning slowly, and the strong wind caused fire to start on the Railway Station. Burning shingles were carried out over the town, and between noon and evening Haileybury was a complete loss.[43]

While residents of Haileybury were impressed by the efficient and compassionate way that Premier Drury directed relief efforts, this goodwill did not extend to Minister Bowman. He was seen as championing the rights of farmers to burn forests which had contributed so greatly to the disaster.[44]

Following the Haileybury inferno, Zavitz obtained Drury's support for amendments to produce a stronger Forest Fires Prevention Act. The changes were spelled out in a detailed memorandum he authored to Bowman. In it, Zavitz explained how, without such changes, "the difficulties surrounding the efforts to detect the origin of fire are considerably emphasized and the owner of the land is not responsible. He should be only to prove that the responsibility does not rest with him, whereas if he is responsible, the Crown, under the suggestion of this Act, is in a position to see that a just penalty is imposed upon him." Zavitz emphasized that the legislation he was seeking was already in effect in New Brunswick, Quebec, and British Columbia.[45]

Despite his empathy for forest-burning northern farmers, Bowman did not block Zavitz's suggested reforms; however, they died on the floor because, as Drury later admitted, he dissolved the legislature "in a fit of temper." This "temper" was in response to a Conservative filibuster to prevent his government from bringing in the transferable vote and a limited degree of proportional representation, which was part of his campaign platform. His regrets at this action, detailed in his memoirs published forty years later, were shaped by the reality that the early dissolution resulted in Howard Ferguson getting the popular plaudits for reforms in forest-fire prevention. These reforms coming down the pipe were ones that he and Zavitz had been working on but were unable to implement fully, because of the hasty election call made six months prior to the normal four-year term.[46]

In 1923 Howard Ferguson, from Kemptville, was elected as Conservative premier and implemented changes to the Forest Fire Prevention Act, changes identical to Zavitz's proposals to Bowman. Thomas Magladery being replaced as the Conservative candidate for Timiskaming during the election campaign helped. His successor, A.J. Kennedy, ran on a platform of the assumed responsibility of landowners for the prevention of forest fires. Kennedy's later interest in the Ontario Municipal Board (OMB), which he chaired with distinction, was distinguished by OMB's role in supervising Haileybury's finances for a decade following the inferno in 1922.

The 1924 amendments to the Forest Fire Prevention Act also enabled Zavitz to summon additional assistance for controlling forest fires and imposed new travel-permission requirements in times of fire dangers.[47] Another measure that gave Ferguson more credit as premier was Zavitz's innovations in the use of air power to combat forest fires. The use of planes had been developed with Bowman's full support during Drury's government and had been introduced when survey planes were flying around Sioux Lookout to examine the spread of White Pine Blister Rust. During that survey, a fire was discovered and a ranger was flown in, as Zavitz recalled, "to take charge" of suppressing the blaze.[48]

Zavitz's experience of the Haileybury disaster further increased his determination to use aircraft as a tool for forest-fire detection and suppression. After reading an article in the *Toronto Star* about the Haileybury fire, which was based on an aerial reconnaissance that compared the blaze to the "tail of the black dragon," he arranged to fly over the inferno to discover this beast for himself. Zavitz was able to sit in an observer's cockpit and make quick drawings of the burning forests below. The pilot guiding Zavitz above the blaze was Roy Maxwell, president of Laurentide Air. Zavitz would subsequently assign the company the task of developing a new division of his Forest Protection Branch, the Ontario Air Service.[49]

Zavitz selected Algonquin Park as the testing ground for using aircraft in fire-fighting. The main base was established at Whitney, with a sub-base at Parry Sound. Fires were reported from aircraft on patrol by dropping messages or landing at a point where a telegraph service or telephones were present. Zavitz concluded:

> The season's operations have clearly demonstrated that for similar country aircraft have no equal for sighting and locating fires. From a height of five or six thousand feet the smoke of a camp fire can be seen for several miles and located within one-quarter mile by an experienced observer. When bad fires occur the chief ranger is able by flying over the area to place his men to greater advantage and when a patrol is finished the officer in charge of the district knows the exact condition of fires throughout the territory. The morale effect on the people within the patrol area is also of great importance.[50]

In addition to detecting fires, Zavitz found the planes excellent at "transporting fire equipment to fires in remote areas, for mapping forest types and taking photographs of particular areas."

During the British Empire Forestry Conference held in Temagami, Ontario, in the summer of 1923, shortly after Ferguson's election as premier, Zavitz highlighted air power as Canada's unique contribution to imperial forestry with a demonstration in Temagami. He arranged for a controlled burn to be set about twenty miles from the Temagami Forest Station, where the imperial foresters were assembled. The delegates stood outside the station, which had a loudspeaker outside. He recalled:

> In a little time we received a call from Bear Island [where the fire tower was located] on the location of the fire about twenty miles away. In a few minutes the plane came over the Station and dropped the same message. Upon receiving the message, the Conference was taken by launch to a point where the fire pump was pumping through the hose, which ran about a quarter of a mile to a canvas tank, which was being used for another pump and hose to reach the fire.[51]

Aircraft soon proved their worth across Ontario. Planes were used for patrolling, surveying, sketching, photography, and the hauling of goods, and were specially designed for the use of the Ontario Air Service. Its headquarters at Sault Ste. Marie provided special training for pilots. Costs for the service were reduced through payment for special transportation services by the Hudson's Bay Company, Indian Treaty supplies, Ontario Hydro, and prospectors. Zavitz recalled how there were "many inspiring examples of devotion to duty" during emergency flights that were carried on during "adverse weather conditions." These emergencies included flights for the conveyance of diphtheria serum to Northern Road camps, doctors to Indian camps, Indians to hospitals, sanitary inspectors to mines, Government officials to urgent business, doctors to summer camps along the rugged Lake Superior coastline, fire fighters, prospectors and woodsmen from the interior to hospitals.[52]

The mix of air power, tougher laws, and changed public attitudes did reduce the extent and devastation caused by forest fires in Northern Ontario. By 1925 Zavitz concluded that the fire danger had been reduced sufficiently so that reforestation could begin in the north. As a history of the Cochrane District published by the Department of Lands and Forests later concluded, "A marked change had taken place in the attitude of the general public toward fire protection, particularly in reporting fires to rangers. This resulted in more fires being extinguished before getting out of control. The lumber companies and the railway companies were also co-operating in every way possible."[53]

The remarkable ability of Zavitz to be good friends with both Drury and Ferguson helped to bring about one of the most important changes in Ontario history — the end of the threat of uncontrolled forest fires caused by human actions. He brought similar stability and gradual improvements to reforestation programs in Southern Ontario, which eventually ended the threat of drought, flooding, and spreading deserts — the consequences of deforestation.

In a speech fourteen years after his electoral defeat by Ferguson, Drury acknowledged Zavitz's careful, informed approach to policy and its implementation. He recalled, "Our successors used this compliment, that they did not repeal one Act we passed or undo one piece of work which we had begun, and these stand today."[54]

— Seven —

A Decade of Environmental Reform

THE CONSERVATIVES HELD POWER in Ontario for eleven years, from the 1923 victory with Howard Ferguson as premier until their defeat in 1934 under his successor, George Henry. During that period, Edmund Zavitz would be their environmental conscience. Using his knowledge, formidable tact, diplomacy, and the lure of genial entertainment such as fishing trips in the forests of Ontario — strategies he would impart to his protege Frank MacDougall — Zavitz persistently prodded reforms out of the provincial government.

The fact that the Tories built upon the reforms that Zavitz and Drury had developed does not mean that they behaved as a United Farmer-Labour government would have (had the outcome of the 1923 general election been different). No labour-influenced government would have seriously contemplated, as Ferguson's did, the use of prison labour in reforestation projects. Neither would have Drury engaged in a tussle with Zavitz over the location of the first reforestation station in Northern Ontario, a battle that delayed the spread of tree planting in the north. Still, in most matters, especially relating to Ontario south of the French River, all political stripes in Ontario had similar policies, that is, until Mitchell Hepburn became leader of the Ontario Liberals in 1929.

Zavitz continued to expand the conservationist role of professional foresters in the north beyond just the mandates of fire protection. His Forest Protection Branch, which in 1928 became the Department of

Forests, pioneered in new technologies. It became a major force in a province where large stretches of isolation made new inventions like aircraft and radio even more important. Radio, first tested by his branch in 1923, became operational in 1927. Originally used for communications on such tasks as the outbreak of forest fires within the Forest Protection Branch, it soon became a radio-telephone system for Northern Ontario. Zavitz now had a tool for sending a barrage of supportive directives from his head office in Toronto to his men in the field, messages that were reinforced by frequent field visits by assistants, such as Arthur H. Richardson.

Zavitz continuously nudged his northern rangers to move beyond the narrow scope of fire protection toward a more comprehensive mission of ecological restoration. Fire rangers were periodically assembled to build morale and update information. One such gathering, billed as an annual meeting, was held in Sudbury from April 7 to April 9, 1927, for the fire rangers of the Sudbury District. Topics presented went beyond fighting fires and into the merits and techniques of reforestation, as discussed by Richardson and Dean C.D. Howe.[1]

While Zavitz's work in the north was helped by one particular backbencher, A.J. Kennedy, he had a number of more powerful backers from the south at the cabinet table. One was John Martin, a friend and neighbour from the riding of Norfolk South who served as minister of agriculture but died in office in 1932. The two men had side-by-side cottages at Turkey Point on Lake Erie. Martin was anxious to have the government take quick action to protect Turkey Point from the dense residential development similar to what had hit Long Point and destroyed its fine wildlife habitat. He supported the recommendation that Turkey Point be a reforestation station under the control of Zavitz. However, it did not become a provincial park until 1956; park rules now prohibit the sort of development that had earlier damaged Long Point.[2]

During the summer of 1923, shortly after his election to the legislature and being made minister of agriculture, Martin told Zavitz that he "wanted to carry out projects needed in Norfolk County." Zavitz's response was that he should "look over the proposal of having the

Government acquire the Normandale plains or desert." Since Martin's knowledge of Turkey Point was limited to the area around the access road they used to reach their cottages, the two men "arranged to meet and look over the proposal, which was done over the next weekend." Next, they

> met the Premier with maps, and he very soon gave me authority to take options on whatever area was required. [The premier and Mrs. Ferguson had previously visited St. Williams and were familiar with the desert-like conditions in the Normandale area.] Mr. Bruce McCall of St. Williams was employed to search options. The Government obtained the area at prices ranging from five to thirty five dollars per acre except on the portion which was still in the Crown, being the old townsite of Charlotteville, chosen by Governor Simcoe in 1795 to become the chief garrison and naval arsenal for Lake Erie.[3]

Zavitz discovered that, in 1795, Lieutenant Governor John Graves Simcoe had written a letter to Lord Dorchester (Sir Guy Carleton), the governor general of British North America, asking him to approve the name of the point because of the former abundance of Wild Turkeys there — "former," because the Wild Turkey had completely disappeared in the area. Simcoe had also documented Fort Norfolk's important role in the War of 1812 and how its ruins were threatened by shifting sand dunes in need of reforestation. Zavitz urged Ferguson to "develop a property of beauty and ability, giving the people of Norfolk and elsewhere a playground on the shores of Lake Erie which will be free to all." He viewed it as the Turkey Point "Historical Park and Forestation."[4]

Turkey Point was put under Zavitz's control as a Reforestation Station, known as Forestry Station No. 2. Zavitz placed the site under the management of his close friend, J.H. White, and in 1927, he supervised the planting of 700,000 trees there. While removing stumps and preparing the ground before reforestation, he was careful to have "healthy specimens of white pine and white oak remain." White took care to leave "clumps or islands of trees for the purpose of natural fertilizer or leaf drop" and to serve as "a wind-break, a fire preventative and furthermore increase the extent of leaf distribution." White also created an arboretum featuring rare local trees, including the Chinquapin Oak. After the sands were stabilized by the reforestation, some areas of Turkey Point seeded itself to oak and other native flora, such as Bird-Foot Violet. The management records kept were so thorough that they were used to aid reforestation throughout Southern Ontario.[5]

Today, in addition to the restoration of Wild Turkey, other extirpated species have returned, including the Bald Eagle and Sandhill Crane. Turkey Point is also a refuge for the Snowy Owl and 117 species of breeding birds. A twenty-hectare nature reserve has been established to protect rare Carolinian tree species, such as the Tulip Tree, Flowering Dogwood, Sassafras, and the Sweet Chestnut. Other rare species include the Southern Flying Squirrel, Hooded Warbler, Dwarf Chinquapin Oak, Green Milkweed, Cylindrical Blazing Star, American Ginseng, and Spotted Wintergreen, and a rare relic spruce bog is protected here. Since 1992 a prescribed burn has been used to protect the rare oak savanna.[6]

Located not far from Turkey Point, the St. Williams Forestry Station No. 1 was developed as a popular tourist attraction by Frank Newman. Picnic tables were supplied, and visitors swam and fished for free in a pond stocked with trout raised in the nearby Normandale hatchery. Churches, inspired by the serenity and beauty of the reforested landscape, held services there. A baseball field was laid out and became the site for many tournaments, bolstered by the field staff's own baseball team. In addition to providing fun activities, the recreational area around the tree nursery became a centre for public education on Ontario's Carolinian zone, making effective use of showy tree species, such as the early-flowering Redbud, a Carolinian tree that has a very limited native range around Lake Erie.[7]

Zavitz was delighted with St. William's success:

> During 1930 the Norfolk Forest Station was the mecca for tourists, exceeding all previous years. The fame of reforestation, as carried on in Ontario, is being broadcast far and wide. American visitors of other years are coming back and bringing their friends who are amazed at the advanced status of forestry in Ontario. All visitors are supplied with experienced guides who are prepared to explain operations in detail and make the tour one of educational value rather than just a matter of sightseeing. Arrangements have, moreover, been made to have interested parties taken care of at Station No. 1 where reclamation, planting, experimental plantations and nursery activities offer an encouraging inspection. Reforestation exhibits were set up at Port Dover, Tillsonburg, Aylmer, St. Thomas, Woodstock, Cloudland and Simcoe ... Several addresses have been given before Kinsmen, Lion and Rotary Clubs and these have been received with interest. The use of lantern-slides, and better still, motion pictures applying directly to reforestation, invariably brings out a larger audience who are given a clearer conception of the project, as illustrated by picture and story.[8]

Earlier, he had reported:

> A comprehensive planting scheme is now in progress ... Through the offices of the reforestation section of the local chamber of commerce, [led by his friend Monroe Landon] options are being procured on abandoned farms suitable for reforestation. These exhibits

were shown last fall at county fairs, and two at flower shows under the Horticultural Society. Requests for additional exhibits in 1929 indicate an increased appreciation of the work being done, while applications for planting material directly traceable to exhibition activities warrant an aggressive continuation."[9]

The projects were not only successful locally but were being appreciated further afield.

Frank Newman encouraged the Norfolk County Council to set even more ambitious goals for reforestation. By the time of Newman's death in 1956, the council had established 809 hectares of County forests, although this was short of his 2,023-hectare target. With the support of the local MLA John Martin during the Conservative governments of Ferguson and Henry, Norfolk established eleven separate county forest tracts. The largest was the Payne Tract, which covered 60 hectares.[10]

In 1933 Zavitz wrestled another reforestation measure into being — the Demonstration Woodlot Program designed for private landowners in Southern Ontario. Administered by the former Midhurst Forestry Station superintendent, Isaac Marritt, it was targeted at areas such as tracts of "privately owned areas of woodland adjacent to a well-travelled road," to educate the public about such benefits as fencing out forests from livestock, a requirement for participation in the program. Signs were placed to show that the forests were supervised by the Forestry Branch, which conducted improvement cuttings, thinnings, and reforesation.[11] Marritt's personal contacts made with area landowners, built up through his former work at the Midhurst Reforestation Station, quickly led to eleven demonstration sites in Simcoe and York Counties. The program, soon popular in Norfolk County, included several sites. One of the current landowners, Harry Barrett, who knew Edmund Zavitz in his retirement years, often walks through his "Norfolk County," a fifty-acre mixed hardwood forest north of Port Dover.[12]

Marritt turned the reforestation station into a provincial park-like setting modelled on the St. Williams Station. In his 1928 annual report, Zavitz explained, "Each year sees the Forest Station used more and more as a community centre. The athletic field and the open-air skating rink draw the young people from far around. In September the Vespa Township school fair was held on the ground and had record attendance." The blossoming of the Midhurst Reforestation Station, which he and Drury had viewed as a desolate wasteland back in 1906, epitomized the redemptive qualities of reforestation in Ontario. Through absorbing what Zavitz termed "natural barnyard manure," mixed with "decomposed humus from swamp land," the site became a powerful symbol of transformation. In his 1931 annual report, Zavitz could truthfully gloat, "People are attracted to this beauty spot and in coming to see it they came in direct contact with the results of reforestation. This is a wonderful object lesson to them. Every means is taken to show them the benefits arising from reforestation as carried out in this province. A show case displaying our bulletins has been placed in the park."

At Midhurst, Marritt effectively pushed municipal reforestation through various public institutions. Some thirty hectares were reforested under his supervision at the Longstaff Prison Farm grounds (north of Toronto) and just under a hectare near the Orillia Provincial Hospital. One of Zavitz's impressive projects, a benchmark in the early protection of the Niagara Escarpment, was the 1928 reforestation of the sixty-three hectare Inglis Falls Municipal Forest, a scenic spot on the Sydenham River just south of Owen Sound. The reforestation projects continued: Markham and Brighton protected their waterworks through reforestation, Beeton's waterworks forest included forty-three hectares, and Coldwater — a community that became characterized by a beautiful stream flowing through its core — planted a twenty-hectare municipal forest. In 1929 Marritt helped the Owen Sound Kiwanis Club to plant a forest of 40,000 trees in one of Owen Sound's parks.[13]

Marritt finalized the work Zavitz had initiated in 1924 when he secured York County's purchase of the 438-hectare Vivian Forest on the Oak Ridges Moraine, through the Agreement Forest Program. Marritt helped the local county council to locate properties, make

Two Billion Trees and Counting

These photos show the dramatic reversal of desertification through the extensive reforestry projects on the Oak Ridges Moraine, undertaken by the Durham Regional Forest. Top image dated 1927, bottom, 1997.

appraisals, and complete the ultimate area to be reforested under the direction of the York County Reforestation Committee. The forest, now named the York Region North Tract, located on the Oak Ridges Moraine in the Municipality of Whitchurch-Stouffville, is a good example of the valuable protection provided by large blocks of forest in Southern Ontario. Birds requiring good forest interior habitat, such as the Red Shouldered Hawk, Goshawk, Pileated Woodpecker, Barred Owl, Tanagers, and colourful warblers all nest here.

The York Region municipal forest comes the closest in the province to realizing Zavitz's ideal of municipal forestry, which was based on the European examples he had studied at Yale. Today, it exceeds 2,033 hectares in eighteen separate forest tracts. As intended by Zavitz, the logging of these forests is restricted to thinnings of the coniferous plantations in order to gradually encourage a transition to a natural mixed-woods forest. A major goal of the municipal forest is to provide passive recreational opportunities such as cross-country skiing, horseback riding, and hiking. Unlike most municipal forests initiated by Zavitz through the Agreement Forest Program, the York Regional Forest is still expanding. In the last fifteen years that the forest has operated without the assistance of the provincial government, York Region has acquired eighty hectares of forests at four different sites with the co-operative assistance of the Nature Conservancy of Canada: the Oak Ridges Moraine Trust, and the provincially funded Oak Ridges Moraine Foundation. The York Region Forestry Department, which manages the tracts, has also become effective in prosecuting private forest owners who illegally clear cut their lands in violation of tree-cutting bylaws.[14]

In 1931 Marritt secured a large-scale Agreement Forest Program in Dufferin County. The county treasurer, James Henderson, had been inspired by the resurrection of healthy forests on the former blowsand deserts at the Midhurst Forestry Station and the Simcoe County Hendrie and Orr Lake Tracts. Henderson initiated a county council motion to purchase a block of 409 hectares of land on Concessions 6 and 8 in Mulmur Township. Eventually, after being expanded to 607 hectares, this block became designated as the main tract of the

A 1923 photo taken by Zavitz shows extensive wastelands in Mulmur Township, Dufferin County.

Dufferin County Forest. It has become an important wildlife habitat for many wildlife species, including White-tailed deer, Ruffed Grouse, Wild Turkey, and porcupine. Another Dufferin tract, the Terra Nova Complex, is a swamp forest and an important deer-wintering yard and trout-spawning area. The Dufferin County Forest is now comprised of 969 hectares. One of its most important acquisitions was the Mono Tract, which buffers an important feature of the Niagara Escarpment: Mono Cliffs Provincial Park.[15] All of these projects were achieved through Zavitz's legacy; without him there may well never have been the Agreement Forest Program or conservation authorities.

Outside of York County, the task of reforesting the Oak Ridges Moraine continued with George Linton, the superintendent of the Orono Reforestation Station. Until 1927, when he was finally able to secure a good supply of water for irrigation, his efforts were challenged by the Herculean task of obtaining adequate water. He found that farm plantings "of impoverished soils and of sand areas ... becoming increasingly popular." Finally, he was having some success in persuading farmers to agree to have their wastelands planted with trees.

In spite of these difficulties, Linton was an effective missionary in spreading the word of reforesting the Oak Ridges Moraine through speaking engagements, letters to the editor of newspapers, ploughing matches, and fall fairs — any event where he might have an opportunity to convert skeptics. He encouraged landowner participation through lectures on reforestation in the Agricultural Short Courses taught in winter months in Clarke and Darlington Townships. Part of the course included slide exhibitions of Zavitz's first plantings in Darlington Township and his early plantings on the farm of C.P. Sisson on the Oak Ridges Moraine. The exhibit also included manufactured articles made from wood obtained from thinnings of Zavitz's pioneer plantings.[16]

One of Zavitz's achievements, following the creation of the Agreement Forest Program in 1921, was to have it adopted by every county government along the Oak Ridges Moraine. This photograph, taken in 1923, shows a step in this program process as the site for the future Northumberland Forest is being selected amidst the bleak sand wasteland. Here, the superintendent of the Orono Reforestation Station, George M. Linton (left), is shown in discussion with county representatives Walter Fowlds and Colonel MacNachton.

From Zavitz's perspective, a hallmark of success regarding the reforestation of Southern Ontario would be having every county government that straddled the Oak Ridges Moraine sign on to the Agreement Forest Program. By 1930 this had been achieved. Linton played a critical role in this, in part by first getting the smaller scale demonstration-plot forests established. The Township of Orono reforested sixty hectares and Bowmanville planted 25,000 trees as a form of unemployment relief during the Depression years.[17] It took Ontario County three years to acquire the land, but, in 1925, it purchased 404 hectares acres in Uxbridge Township on a piece of moraine land that was "mostly poor sand land with occasional pieces of woodland." One of the most spectacular of the Moraine success stories was the Northumberland Forest.

Ontario County had been talking about reforestation since 1910, but Linton finally prodded them into action. While the initial purchase was "the bare minimum for the program of 1,000 acres," over time this expanded to 2,023 hectares. The ecological benefits of the Northumberland Forest are concentrated on one large swath of the Oak Ridges Moraine where the massive reforestation helped to bring Cobourg Creek back to life, restoring Brook Trout populations and later the Atlantic Salmon.[18]

In Eastern Ontario, Zavitz's success in reforestation was largely due to the vision and determination of a remarkable personality, the agronomist Ferdinand Larose. Before he came on the scene, Zavitz lacked solid contacts in this part of the province, an area largely ignored in his 1908 wasteland report. Larose entered government service in 1921, during Drury's administration, as the first French-speaking agronomist employed by the Ontario government. He soon became Zavitz's man in Eastern Ontario.

Like other boosters of Zavitz's reforestation program, Larose got things going with the same demonstration plots carried out by municipalities. He originally stimulated reforestation to protect the water supply of his home community of Plantagenet, located in Russell County, east of Ottawa, and started similar projects in Bourget and

Clarence Creek. In his role as an Ontario government agricultural extension agent, Larose also persuaded farmers to reforest lands to protect their own wells and sparked the beginning of demonstration forests in Glengarry, Stormont, and Dundas Counties, the largest being an 81-hectare forest in Glengarry near the village of Maxville.

In his support of reforestation, Larose was similar to many of the other government-based agricultural extension agents throughout Ontario. He combined his advocacy with a great sense of belief in his mission. One of his tasks was to help farmers develop organization muscle through the nurturing of the Ontario branch of the Quebec-based League of Catholic Farmers.[19] His training in agronomy in Quebec provided a useful foundation for his task of restoring forest cover to Eastern Ontario. He had obtained his Bachelor of Science in Agriculture at Laval University's

The Larose Forest, located in Russell County right on the edge of Ottawa's Greenbelt, is part of the agreement forests in Eastern Ontario. It is shown rising from the sand wastes that still lie around its edges. The forest was named in honour of Ferdinand Larose, an agronomist working with the Ontario Department of Agriculture. He is also known as the driving force behind the reforestation of Eastern Ontario. Photo taken in 1963.

campus in Oka, a community whose blowing deserts were really the birthplace of reforestation in Canada. Prior to the Catholic Church and the Quebec government undertaking reforestation here, the sandy soils around Oka were similar to the wasteland east of Ottawa known as the Bourget Desert. Streams had dried up in the summer months, massive spring floods were taking place annually (reducing the growing season), and the shifting sands were threatening more farms with soil erosion. Larose had formidable obstacles to face, since Russell County, adjacent to the City of Ottawa, had only 4 percent of its rural landscape in forest cover.[20]

Much like Zavitz, Larose had a deep interest in Ontario history, which helped shape his reforestation efforts. Frequently, he would choose a site for his restored forests because it had once been the site of exceptional White Pine stands in Eastern Ontario. Much like the American Pinchot, he condemned the powerful families responsible for creating a desert-like environment. One particular family was in the drainage business and was obsessed with draining the Alfred Bog in Russell County. Larose deplored how drainage schemes "have considerably lowered the water level in the streams and in the surrounding soils, with ill effects in springs and wells and in the growing of crops, especially in the years of low precipitation." He documented how schemes to burn off peat to farm the clay underneath resulted in massive forest fires, sometimes so severe as to force expensive repairs to roadways.[21] In describing conditions in Eastern Ontario, Larose asked, "Is there any need for Conservation? We have floods, soil erosion, drought and our land is denuded of trees. Where is our timber wealth? Where is our lumber industry? What has happened to our numerous saw mills? They have disappeared with the woods. What more can we say."[22]

Zavitz's 1911 reforestation effort at the estate of Senator W.C. Edwards in Rockland was a boon to Larose. He could take non-believing skeptics to a site where they could see how new forests could bring back desert-like land from the dead. The Red Pines here tended to show off "very nicely" to the degree that the young forest was "considered to be one of the best, if not the best, in the Province of Ontario."[23] Larose stirred the public imagination over reforestation in Eastern Ontario through a deluge of other activities designed to promote the welfare of farmers. In 1926, for instance, his "general campaign" for reforestation and soil conservation was combined with efforts to encourage the growing of small fruits and vegetables for home canning, an initiative that was taken up by 135 families. He also helped to distribute seedlings to children for planting on their home farms to create windbreaks and shelterbelts.

After six years of extension work, in 1927 Larose finally persuaded the United Counties of Prescott and Russell to sign the Agreement Forest Program with the province and purchase 446 hectares of land. This was eventually expanded into about 180 square kilometres. Both tree planting and land purchases continued for many years after his death, with the full support of the United Counties Council. Although most of the land acquired was drifting sand, there were some small patches of aspen and elm to be protected. Natural regeneration was encouraged by protective measures such as fire suppression and fencing from cattle. Away from the forest's core, Larose undertook the purchase of the Moose Creek Bog. This 2,032-hectare southern edge of the Larose Forest had been intensively drained and both its peat bog and swamp forests heavily burned. At the time of its purchase, it was "nothing except bare peat and muck ... and all that [remained were] isolated patches of tamarack."[24]

Although the great restored forest was named in his honour, Larose, like Zavitz, did his best to share the credit by singling out Marshall Rathwell, who had served on the elected Prescott and Russell County Council for eighteen years. Throughout, he had functioned as chair of the County's Forestry Committee, while Larose was secretary-treasurer.

Planting of the Larose Forest began in 1928, with Red Pine seedlings set in the 40.5 hectares of blow-sand land. White Pine and White Spruce were added later. By the 1940s and 1950s, planting in the Larose Forest was at the rate of one million trees per year. When planting levels were at 200,000 trees annually in the 1970s, it was possible to include poplar, birch, and other deciduous trees, since the desert-like conditions that made their survival difficult had improved considerably. Most of the Larose Forest now resembles a natural woodland. Beavers, which were

The Honourable René Brunelle, Ontario minister of lands and forests, chats with Mrs. Larose, widow of Ferdinand Larose, beside the newly dedicated plaque mounted on a six-ton boulder.

originally absent when the area was a degraded desert, have returned and through their flooding efforts have created significant wetlands. This, in turn, provides an important wildlife habitat for wetland dependent birds such as Wood Ducks, Sora, Virginia Rail, and the American Bittern. The large extent of forest cover has made it one of the last nesting stands in Ontario for the Evening Grosbeak. Other species that have also returned include the River Otter, Fisher, Porcupine, Red-Shouldered Hawk, and Red-headed and Pileated Woodpeckers. While numbers are still low, Black Bear populations are also recovering. The forest has become an important refuge for the endangered Blanding's Turtle, and its vernal pools support several salamander species.

The Larose Forest is now a core habitat for Eastern Ontario's moose population. A herd of about three hundred roam between Larose Forest, Alfred Bog, and the Ottawa Greenbelt. Although moose populations are controlled by an annual hunt, it is still one of the most densely populated moose herds in the world, and the only one whose habitat is largely surrounded by agricultural land, largely dairy farms. The herd had crossed the Ottawa River from the Laurentian Shield in Quebec and were first discovered when some were shot by hunters in 1985.[25] For wildlife populations and the interested public, the creation of the Larose Forest was an extraordinary example of how a forest could be restored. The beavers, that have done so much to shape the forest in Larose's lifetime, had vanished by 1825. Studies found that "very few" ducks of any description were breeding in the lands when it was purchased. While Larose once decried the fact that the "larger and more spectacular forms of wildlife" of interest to the average citizen had completely disappeared, the forest named after him today provides an extensive refuge for such animals. The Larose Forest is yet another example of a positive outcome stemming from the Agreement Forest Program Zavitz had developed along with Drury.

While Larose was working in Russell County, Zavitz was also taking action to restore and protect forests in other parts of Eastern Ontario, which, unlike the reforestation projects on arable land sponsored by Larose, were located on the rocky Canadian Shield. His Forest Protection Branch's main zone, known as the Tweed District, was in the vicinity of the community of Tweed in Hastings County. Some thirty plantations of several species of trees were planted here during the 1920s. While records were kept of the establishment of these plantings, there was insufficient technical staff to carry out inspections and monitor success. Plans for reforestation based on the planting of native seed trees in areas that were heavily logged also went nowhere because of lack of resources.[26]

Isaac Marritt's Demonstration Woodlots Program for private landowners sought to return pines to the Canadian Shield on private lands where the coniferous species had been stripped away by earlier logging. J.H. White explained to Zavitz on the eve of obtaining funding for this program:

> [This was a] region of more or less derelict hardwood following the extraction of the softwoods. A very large

proportion of the trees is privately owned, but the woodlots at present have little value beyond the hardwood market, and much of the cleared land has low agricultural value. It would seem that a good opportunity exists to raise the productive value of this large property and improve the economic status of the owners. Woodland of this type can be rehabilitated most cheaply by planting conifers. The proposition would be similar to what has been carried on for the last 20 years to the south. If small demonstration plantations were established adjacent to settlements, followed by a definite campaign in the winter by district officers, one could expect at least as good a response as has arisen in other parts of the province. An influence towards more care with forest fires could also be expected.[27]

Getting the Demonstration Forest Program started on the Canadian Shield was slow going. In the program's second year, however, a few sites were established in Renfrew and Frontenac Counties. These reforested plots were part of a lengthy process of educating landowners on the benefits of reintroducing conifers to land that had been stripped in the past by loggers who refused to leave seed trees.[28]

White's call for the reintroduction of conifers was boldly taken up by the Forest Protection Branch's district forester for Parry Sound, Peter McEwen, a recent graduate of the University of Toronto's Forestry program. Like his mentor, Edmund Zavitz, McEwen had a flair for turning wasteland into healthy forest through a dramatic public demonstration project. He chose a hill stripped bare of trees, close to what became the district headquarters of Zavitz's Department of Forests in 1927. It was to be the site of a twenty-seven-metre-tall steel forest-fire detection tower. Unlike most such towers, McEwen had it carefully designed with safety features, such as landings that also served as observation posts; these special features made it a tourist magnet, and the tower attracted thousands of visitors annually.

McEwen turned the barren wasteland below the tower into a dramatic demonstration of reforestation, with the intention of educating the public. In 1927 he took the first step of planting 15,000 Red, White, and Jack Pines on six hectares around the base of the tower. These trees were interspersed by a few native maple and White Birch trees. In 1930 this was complemented by a larger seven-hectare planting of mixed coniferous and hardwood forest. To build morale among his staff, McEwen had all Forest Department employees in his district involved in tree planting, including clerks and forest-fire prevention rangers. It was said that there were "never any idle men sitting around waiting for fires at the district headquarters."[29]

To attract more visitors, McEwen followed the St. Williams' model of providing recreational facilities on the site. He had workers with a team of horses scour out a fish pond and line it with stone. A rock garden was created by having a scow haul in decorative rocks from Georgian Bay, and gravel and planting soil were brought in to support attractive flower beds. The resulting beauty spot, fishing pond, and attraction for tourists was a great success, contributing much to public education on the merits of forest protection and landscape enhancement.

Zavitz continued to rely heavily on James White for advice on Northern Ontario. He identified the areas where reforestation could be started once stronger regulations and air power had sufficiently suppressed forest-fire dangers. White pointed out, "The Kirkwood plains [east of Sault Ste. Marie] offer an opportunity for experimental reforestation plots hard to duplicate."[30] By using the term "hard to duplicate," White meant that nowhere else in the north were there conditions resembling the blow-sand disaster caused by fire, too-intensive logging, and ill-considered agriculture of the Kirkwood region. For Zavitz this was the Kirkwood "desert."[31]

White's call for action at Kirkwood was reinforced in 1928 when Zavitz, as deputy minister of forests, sent out a circular to district foresters, asking for "suitable areas for demonstration planting" in the north. He later recalled, "In reply to the circular, F.A. MacDougall, the District Forester at Sault Ste. Marie, suggested the Kirkwood Desert.

Kirkwood is a township north of Thessalon and was one of several townships where the forest had largely disappeared. Planting was begun in 1928, and by 1943, 6,000 acres had been restocked."[32]

As Zavitz indicated, lessons learned at Kirkwood were later applied to other parts of the Canadian Shield. He noted that following the 1943 evaluation of the first plantings, the program was expanded. In November 1943, an additional 14,000 acres were planted and set aside as the Kirkwood Forest Management Unit, eventually to be used in "Crown land planting for demonstration purposes ... in Parry Sound, North Bay, and Pembroke district."[33]

In 1929 Zavitz secured another victory with the establishment of a system of provincial forests. New legislation abolished the old designation of Provincial Forest Reserves and turned those in Temagami, Sibley (later Sleeping Giant Provincial Park), Eastern (in Hastings County), and Algoma Reserves into provincial forests. In the past, these designations meant that no land could be sold for agriculture, and fire controls barred admission to all but logging operations. While the ban on farming remained, the new Provincial Forests now welcomed tourists, another sign of the reduction of northern fire hazards. Provincial Forests were to become a "sportsmen's paradise" for activities such as nature appreciation, camping, canoeing, hunting, and fishing. The former Temagami Reserve was expanded to include the township of South Lorrain and the Gillies Lumber Limit, north of the Temagami village, to preserve the "scenic beauty of Temagami Lake." Three new Provincial Forests were also established. The one at Georgian Bay near Parry Sound was now administered by Peter McEwen; another, Wanapitei, was in effect a western extension of the Temagami Provincial Forest. The third was the Kawartha Provincial Forest, located in the Canadian Shield-Trent waterway system. One of the most important aspects of this legislation was that Zavitz's Department of Forests now had a land base to administer, outside of Algonquin Park. District foresters, all professional foresters educated at the University of Toronto, would now administer these lands according to principles of sustained-yield forestry. These lands were no longer the jurisdiction of Crown timber agents, who were heavily influenced by local logging interests.

White was elated: "The Provincial Forests Act of 1929 specified nearly 15,000,000 acres [6,070,284 hectares] as coming under its provisions. The important thing about these areas in contrast to their former status as Forest Reserves lies in the placing of their administration in the hands of a forester and implying administration with recognition of the interests of the future." He understood that while restrictions in management were imposed by existing leases of timber, that "practically all" of the these provincial forests contained the opportunity "to direct steps in management now."[34]

Zavitz's Department of Forests employees demonstrated considerable zeal in their quest to ecologically restore the Provincial Forests. Some of the most aggressive action taken in this regard was in the area studied by White and his team of foresters, over the period of 1929 to1934, for the Commission of Conservation. The area included the Trent Watershed and the Eastern and Kawartha Provincial Forests. Here, farming continued to encroach illegally on what should have been reserved Crown-land Provincial Forests. With Zavitz's encouragement, his dedicated staff dug out "many lots of questionable ownership ... from the official record." For the Eastern Reserve, a "second list of Crown lots by concession and township" was prepared. Investigations on fraud were carried out where records showed that no taxes had been paid, and these cases were subsequently checked by "field investigations" and title searches. Forester Lester H. Crosbie found that frauds caused the Crown "not to gain a new settler," but to lose "its timber with no return to show for it." He found that many lands seized by municipalities for non-payment of taxes were actually Crown land. He complained to J.F. Sharpe, the administrator of Provincial Forests, that "maladminstration or insufficient administration of crown land regulation" was a serious problem. He denounced municipalities for "taxing and selling for back taxes crown lands" in areas that were "potentially forest land and unsuitable for agricultural purposes."[35] Studies of the Kawartha Forest confirmed White's earlier research of the devastation to forest soils by past logging. Surveys discovered that a quarter of its 138 square miles [279 square kilometres] had been degraded to "barren-or semi-barren lands."[36]

Records of the administration of provincial forests during the five-year period that Zavitz and his dedicated team of foresters controlled them (1929–34) show a clear pattern of zealous administration of these lands in the interests of ecological protection. Quick action was taken on contentious issues such as illegal land sales, the removal of fire-prone logging slash, and prospectors attempting to seize Crown land through the manipulation of squatters' rights and exploration permits. The biggest confrontation took place in the Georgian Bay Provincial Forest between Peter McEwen and the Marathon Logging Company, who were logging illegally. McEwen reported to Zavitz in a November 8, 1931, memorandum: "The trespass looks like a deliberate steal by the Company of timber which they wanted and which they were determined to get even after their application had been refused; and their methods show a callous indifference towards all regulations." He urged that "a precedent of strict disciplinary measures in cases of infringement and trespass in a Provincial Forest be set." Zavitz backed McEwen's request for prosecution of Marathon for its illegal cutting of 2,000 Red Pine trees.[37]

While Zavitz's experiences had demonstrated the importance of aircraft in detecting violators of conservation regulations, it was his skilled protege, Frank A. MacDougall, who extended the use of airplanes in forestry. When he was the district forester posted in Sault Ste. Marie and planning the strategies for northern reforestation, he also learned to fly at the base of the Provincial Air Service headquartered in the city.[38] Three years later, in 1931, he was appointed superintendent of Algonquin Provincial Park, where he became known as the "flying superintendent." MacDougall's favourite plane for patrol work, which he undertook personally, was the Fairchild KR-34. This plane had been specifically developed by the Department of Forests for law enforcement purposes. The Fairchild patrols were especially effective in winter, when poachers' paths could be easily detected in the snow.

— A Decade of Environmental Reform —

Frank Archibald MacDougall was a critical figure in realizing Zavitz's vision to bring the Imperial ideas of forest conservation to Ontario. A veteran of the First World War and a survivor of the gas attack at Vimy Ridge, he, like most of the team of talented foresters hired by Zavitz, was a graduate of the University of Toronto's Faculty of Forestry. He quickly won distinction through his reforestation work in Northern Ontario and performing aircraft patrols to suppress beaver poaching. In 1931 he became the "flying superintendent" of Algonquin Park. While serving as deputy minister of forests, he brought Ontario's forests under conservation management through land-use planning of Crown lands, the creation of shoreline reservations to protect watercourses, and extensive expansion of Ontario's network of provincial parks.

Some sixteen successful convictions for poaching were made annually after air patrols were launched. It was said that MacDougall "put the fear of the Lord —almost literally as well as figuratively — into veteran poachers who had been reaping a tidy harvest in the park for some time."[39]

In accordance with Zavitz's expectations, MacDougall administered Algonquin according to the ideals of the imperial forestry developed in India — practices characterized by a great concern for the protection of species diversity. He developed park zoning to suppress cottage development and define where logging was prohibited. Efforts to exterminate wolves ended. New restrictions were placed on fishing, closing lakes where biologists' studies had reported that species were at risk, and nature interpretation programs were launched for visitors.[40]

The pattern of incremental reform, however, crashed with the election of the Liberals under the leadership of Mitchell Hepburn in 1934. He sabotaged the system that Zavitz had developed to ensure that foresters had control over some major stretches of Crown land in Northern Ontario through two clever measures: Hepburn removed Zavitz as deputy minister of forests and replaced him with Frederick Noad, a "veteran lumberman" who was living in Toronto at this time and working as a journalist for *Saturday Night*. Noad promptly fired most of the foresters responsible for the administration of provincial forests and replaced them with Liberal Party political hacks, most with limited experience pertaining to forest administration. Problems were also accelerated by the firing of game wardens. Poaching during the Hepburn era became so entrenched that moose, caribou, and bear populations were in danger of collapsing. Fortunately, a decade after the reorganization of the Department of Lands and Forest in 1941, wildlife populations had recovered. Records in the Provincial Archives on the administration of provincial forests abruptly cease with these events.

Before becoming deputy minister, Noad was a political journalist — a cog in the Liberal Party propaganda machine. It seems he developed the idea of firing foresters after meeting disgruntled Crown timber agents during a tour of Northern Ontario, before becoming deputy minister of forests in place of Zavitz. As a lumberman, he was opposed to conservationist regulations enforced by foresters. In his youth, Noad had been a "cant-dog man," a term then used to describe logging rivermen. With his connections to logging, he rose through the ranks, supported by logging barons with ties to the Liberal Party. During the period that the Liberals were in power, it was customary for them to formally nominate the Crown timber agents.[41]

For Zavitz the trauma of these events amounted to a second exile to agriculture, but the shock was worse because it came at a time of profound personal crisis. In 1933 the eldest son of Jessie and Edmund Zavitz, John Dryden, died at their family farm in Norfolk County at the age of twenty-seven. The son's suicide was related to mental illness, which he had struggled with for many years. He had made a marriage proposal, which was rejected, and then took his life. His younger brother Dean found the body in an isolated part of the Forestville farm. For years, the Zavitz family had coped to the best of their ability without any outside help, and like so many others dealing with mental issues, even today, lost their child to suicide. This tragic loss was followed by Jessie's death two years later on August 11, 1935.[42]

A number of personal factors enabled Zavitz to cope with the combination of personal and professional trauma. His always supportive mother, Dorothy, helped him through the crisis. She would live to the remarkable age of ninety-six, dying in 1941, by which time Zavitz had reversed Hepburn's policy thrusts. His equally supportive stepfather, I.L. Pond, enjoyed a similar longevity, living until 1939. Zavitz was also strengthened by the reality that he was revered as a hero throughout rural Ontario, particularly in areas previously devastated by the spreading deserts, deserts that he had conquered. The public dismay emanating from Zavitz's removal as deputy minister eventually came to the attention of Mitchell Hepburn, who despite his eccentricities, rich cronies, "zig-zags," and policy reversals, was still, like Drury, a farmer-premier.

— Eight —

From Disaster to Triumph

During the Great Depression, Ontario experienced many horrors similar to those so vividly illustrated by the Dust Bowl of the Great Western Plains — all the result of the rampant degradation of the natural world. With the province's forestry regulations now weakened, once again massive forest fires began to incinerate human communities. Once again, deserts were spreading over once-fertile lands, calling to mind the spectre of a Biblical plague cursing the descendants of the pioneers who sold their forest heritage for a bag of potash. Farmers in a panic scrambled for water for their livestock as previously reliable streams, springs, and wells dried up in the summer heat. The same rural residents would find themselves evacuated from their farms by torrential springtime floods.

Eventually, in response to these disasters in Ontario, public policies would emerge resembling the conservationist agenda of the New Deal fostered in the United States by the American President Franklin Roosevelt in 1933–45. However, the circumstances in which the new public polices emerged on both sides of the border were vastly different. In the United States, these were developed by Roosevelt at the suggestion of conservationist friends and advisors. Of these, the most notable were Gifford Pinchot, now governor of Pennsylvania; and Richard St. Barbe Baker, who met with Roosevelt in 1932 before his election. In Ontario, however, new policies emerged from the determination of Edmund Zavitz, working primarily with a handful of like-minded friends.

The disaster facing Zavitz was not just that of the plague of fire, floods, and dust storms. He was also facing personal and political disaster with much public opinion on the side of his political enemies, largely due to agitation being encouraged by Frederick Noad. Hostility to conservation was especially intense in Northern Ontario, where the catastrophes of fire that just over a decade ago had burnt towns and claimed lives had now been largely forgotten. The Liberals swept every northern seat, even that held by A.J. Kennedy, whose career was closely linked to promoting the public good.

The death of Zavitz's wife, his first-born son, and the persecution of his colleagues in the Forest Department he had built up since 1912 took any suggestion of humour out of his dramatic photographs of Ontario's landscapes. Previously, he could joke at people standing on the giant pine stumps in the desert or lying over logging debris. For the

In the midst of his conflicts with the Mitch Hepburn government, Zavitz took this evocative 1935 photograph of a graveyard in Grey County being destroyed by wind-blown sand, photo dated 1935.

first time, some of Zavitz's tragedy appears in his photographs. This is seen most vividly in the image of a graveyard in Grey County, where the wind-whipped, sand-covered cemetery appears on the verge of having its coffins ripped out of the earth.

The forestry people knew it was Noad, not Zavitz, who was behind their firings (Zavitz would help to get them re-hired after the Liberals lost power); Zavitz himself had been unceremoniously removed from the position in the Department of Forestry he had created, transferred without loss of pay to the role of chief of reforestation (which had been the position held by assistant A.H. Richardson) but removed from its control, and confined to activities in Southern Ontario. Richardson, a good friend, would remain as an assistant. Though this demotion was hard on Zavitz, it was even tougher for the department foresters, many of whom were outright fired. The father-in-law of Dolf Wynia, Al Barnes, who eventually became the general manager of the Conservation Authorities Branch, was informed of his firing while engaged in the

Edmund Zavitz also took the picture of the church next to the cemetery in Grey County, an indication that it too was at risk of being buried in sand.

reforestation of Camp Borden. Foresters who had lived through this era were interviewed for the 1967 centennial history of the Department of Lands and Forests. All described it as a "traumatic experience."[1]

Noad's purges and budget slashing contributed to the last forest fire in Ontario that killed people other than those employed to suppress it. With the removal of the requirement for farmers to have permits to burn, burning became unregulated, setting the stage for disaster. Under Zavitz's watch, the Forest Department would remove fire-prone slash from logging areas when it was considered to be a hazard. Under cutbacks initiated by Noad, this procedure stopped. Purchases of fire-fighting equipment also declined. In northwestern Ontario, the Township of Dance in the District of Fort Francis wrested such an exemption from fire permits in 1936, once more giving farmers the right to burn forests at any time. The same pattern as that experienced by Haileybury followed. Numerous settler-clearing fires were combined by a wind into a single enormous blaze, which burnt up 37,231 hectares on October 10, 1937, and killed twenty people.[2]

Zavitz's investigations into the forest-killing effects of air pollution from metal refineries died in Liberal budget cuts. Speaking to a parliamentary committee in 1940, Zavitz had pleaded with them to conduct the needed scientific inquiry on the impact of air pollution on forests. He pointed out, "In the old days before they put up those 500 foot towers [stacks] up, the local areas were all affected. Now that the fumes are carried over larger areas, they are possibly disseminated a great deal and that is the reason it is showing up ... at large distances from Sudbury."[3] Noad and his sponsors, notably the then-minister of lands and forests Peter Heenan, denied the importance of scientific study to protecting forests. Their dissenting attitudes were supported by an army of Crown timber agents, now reappointed to their former forest fire-fighting duties on the recommendations of logging interests.

Noad accepted advice from these logging interests to develop a formula by which "hatchet men" were employed to fire professional foresters. Peter McEwen, a forester who stood up to illegal loggers, received such a visit. The morning after the Liberal sweep, a local party boss came to fire him. McEwen, however, was in the bush at the time.

Noad did manage to fire fifteen foresters and get rid of another five through such means as provoking early retirement. Two determined foresters, Keith Atcheson and J.A. Brodie, refused to leave their posts. They received their back pay when Noad was fired by Hepburn after only ten months in office.[4]

Zavitz was a close friend of Noad's only public critic, J.C.W. Irwin, the two having been involved in the creation of the Southern Ontario section of the Professional Foresters Association. Irwin, a Toronto book publisher, was a partner in the firm Clarke, Irwin & Company, founded in 1930. According to Irwin, they went into book publishing to strengthen public support for forest protection and launched the conservationist group Save Ontario's Forests. A graduate of the Toronto School of Forestry, Irwin represented the faculty on the university's senate. He successfully fought off efforts by the deputy minister of lands and forests, Walter Cain, who was co-operating in Noad's purges, to have him expelled from the forestry profession.

A joint denunciation of Irwin by the Liberals and Conservatives followed a speech he gave at the University of Toronto in February 1943, in which he "fingered Ontario as being the worst offender" in the governments across Canada for refusal "to implement genuine forestry reform because of their desire to retain control over the woodlands for patronage purposes."[5] The *Globe and Mail* reported that Irwin was "roundly applauded by a big audience." It is difficult to determine the impact of this condemnation. It may have led to friction between John Irwin and his partners in Clarke Irwin, since shortly afterwards Irwin broke with them and founded his own publishing company, the Book Society of Canada. However, it was generally said that the dispute was private and not publicly commented upon at the time. Forest conservationists from across Canada had great respect for Irwin's work, as demonstrated by the Saskatchewan CCF government's decision to appoint him to an advisory committee on the province's forests in 1944.[6] Zavitz had provided Irwin information used to expose the harm being done to Ontario forests, and Irwin used this information to publicly condemn Hepburn's government.

Adding to the building controversy over Noad, the Quebec government — well aware of its province's vulnerability to Ontario forest fires — was once again protesting Ontario's stance on forestry measures. Noad's reign would be short-lived, even though he was strongly backed by minister Heenan. In 1935 Mitch Hepburn clashed with Noad over the proposed firing of more foresters while Minister Peter Heenan was out of the country on a personal business trip. Hepburn won. Noad's sudden departure put Zavitz in a position to launch a campaign, which a decade later would result in the creation of Ontario's conservation authorities. In previous years, he had not been able to launch such an initiative, his time being taken up with the difficult task of supporting his foresters in conflict with illegal loggers and others encroaching on Crown land for devious reasons.

During the Great Depression, after an onslaught of drought followed by devastating floods in Ontario, Zavitz's former student and long-time friend Watson Porter (right) led the campaign to create conservation authorities through the Ontario Conservation and Reforestation Association, which he had founded by in 1937. Here, Porter, who protected the forests on his own farm, is being given a certificate by the Department of Lands and Forests representative, R. Thompson, indicating that his land was recognized as a model tree farm by the Department of Lands and Forests.

In 1936, there were a number of conditions in Zavitz's favour. Of these, the most important was that he could involve a number of close friends in renewing the call for protection of Ontario's forests. Ernest Charles Drury had rejoined the Liberal Party and Hepburn awarded him with the position of sheriff of Simcoe County. While many saw this as a minor position, for Drury it had the benefit of putting him inside the county administration building, where he could lobby more effectively for more forests. As well, the floods and drought of these years impacted on another close friend, a former OAC student of his, Watson Porter. Zavitz recalled how Porter "rendered valuable assistance to awakening the public to the necessity of proper land use, especially the reforesting of watersheds and waste lands." He particularly praised Porter's "enthusiastic leadership" in the cause of reforestation. In 1937, Monroe Landon, E.C. Drury, and Watson Porter had been key founders of the Ontario Conservation and Reforestation Association (OCRA), which, Zavitz recalled, proved "a vital factor in arousing public opinion."[7]

While Porter had learned the relationship between forest cover and watershed protection from Zavitz in the OAC classroom, it was the experience of exceptional drought and floods on his farm near London, Ontario, that led to his becoming more involved. He observed how it was "pitiful to see cattle milling around dried-up water holes" because "wells that had never failed before went dry." Following drought came the great Thames River flood, requiring the evacuation of the west end of London. Porter, owner of *The Farmer's Advocate*, wrote, "Something must be wrong when farmers are obliged to draw water in the summer and must be rescued in life-boats from their upstairs windows in the winter." He saw the Oak Ridges Moraine as "practically valueless for farming purposes but of considerable natural importance to the farms lying between those light ridges and the lake." Streams following from the moraine were "dried up in summer," but "taking out our bridges, causing serious erosion of the soil" in the spring.[8]

Porter's protests helped launch reforestation in his home county of Middlesex. Here, proposals from the county engineer had been blocked by a slim majority of councillors until the Thames flooding disaster. The county eventually purchased 97 hectares to reforest part of the Thames' headwaters. Zavitz marked the occasion in a May 13, 1938, speech, where he indicated the project would primarily use deciduous trees, unlike the past reliance on conifers. A press release recorded his remarks: "'Much fine timber has been taken out of Middlesex in the past, but at the present time all Southern Ontario is short of supplies of many hardwoods. The growth and development of Middlesex County owned forest will be followed with a great deal of interest,' claimed Mr. Zavitz."[9]

The Middlesex County Forest eventually covered 809 hectares around a major Thames tributary. Now known as the Mosa-Bradshaw Forest, or sometimes as Skunk's Misery, it has become the largest forest complex in southwestern Ontario. The Lower Thames Conservation Authority acquired another 121 hectares to expand the county's holdings. The forest provides a refuge for twenty-four species at risk, including the Acadian Flycatcher and Hooded Warbler. Horse-logging operations, paid for by a local Presbyterian parish, maintain sustainable practices.[10]

Drury, Porter, Landon, and Zavitz were involved in the launch of the newly created Ontario Conservation and Reforestation Association, held in Barrie from August 11 to 17, 1937. Most of the event was focused on honouring the work of Drury and Nelson Monteith and the unveiling of a cairn to celebrate the Hendrie Forest as the first county forest in Ontario. Delegates toured Simcoe County, visiting all its reforestation projects and likely those sand deserts still requiring attention. What they did was the basic model for the subsequent Field Days held by the OCRA.

At the banquet held in his honour, Drury began by expressing thanks to his friend, Mr. Zavitz. He used Biblical examples of the travails of the Israelites in the wilderness to describe Ontario's experiences throughout the ecological devastation unleashed during the Great Depression. Drury explained, "In our own time, the Grand and the Thames Rivers, through no other reason than the over-clearing of their watersheds, have become streams that instead of being as they were at first, an inestimable benefit to the communities along their course ... have become serious menaces, at one time being the source of destructive floods, at another time having so little water that they became polluted and stagnant." He warned that much of Ontario was threatened by the spread of "Saharas on a small scale," although reforestation in Simcoe County had reversed such threats by the growth of trees that now "give off thousands of tons of water."[11]

While Watson Porter was the public front man, Zavitz provided the organizational muscle for OCRA through his leadership of the district foresters. In 1934 Zavitz had Southern Ontario divided into six zones, each with its own forester. Their basic job was to encourage more reforestation through any means they could devise, including use of publicity, booths at fairs, and events designed to raise awareness of existing government programs for reforestation. It is interesting to note that the various OCRA chapter areas corresponded to these zones. Eastern Ontario, for instance, was Zone 6.

Within these zones, district foresters developed the OCRA-sponsored Field Day activities, supplied vehicles and equipment, and distributed booklets to people who showed up for the events. Free tours,

Springbank Drive in London, 1937. The devastating Thames River Flood of 1937 was the critical event that propelled the birth of conservation authorities. Watson Porter, as a student of Zavitz at OAC, understood the relationship between deforestation and flooding. However, it was not until he saw the impact of this flooding personally that he persistently campaigned for the creation of the authorities.

viewed as opportunities to educate the public on the importance of reforestation, were held at the reforestation stations and often involved a full day and hundreds of miles of travel. Along the way, participants examined reforestation projects and samples of pollution problems. Participants were provided with a mimeographed book of fifty pages containing many illustrations of local environmental problems as well as the efforts being made to address them through reforestation. The Field Days were timed to coincide with county reforestation committees' annual reports, which often included announcements of new land purchases. Many of the activities were assisted by the county forest custodians living on site. These wardens were dedicated and highly motivated, many remaining at their posts until their retirement in the 1980s. Often, modified Field Days were held for children, and these featured guided talks and walks taken through the woods.[12]

Watson Porter regularly reported on Field Days through *The Farmers' Advocate*. One of his favourite sources of interesting quotations was E.C. Drury, who by now had assumed some of the thundering oration skills of an Old Testament prophet. When Drury explained that it only took a decade for a planted forest to protect watersheds, he frequently expressed conservation deeds in Christian terms: the "fruit of repentance" for human "sins against nature."[13]

One of the most dramatic acts of "redemption" came as a result of the first OCRA Eastern Ontario Field Days, which resurrected the slowly moving reforestation efforts there. The September 20, 1938, event was hosted by Marshall Rathwell, reeve of Cumberland Township, and involved a tour of the Larose Forest. Although the plantings here were on a large scale, the trees were still quite young. What really drove home the reforestation message was the tour of the fourteen-hectare pine forest Zavitz had planted for Senator Edwards in 1911. The leading regional newspaper, the *Ottawa Citizen*, recorded Porter's words: "The field day was a revelation ... 'The 20-hectare lot in Rockland,' he said, 'was without peer in all Ontario. The forests we have seen should be just symbols of what every farmers should be doing on his own land,' Mr Porter said, in urging all to become 'militant apostles of reforestation and conservation.'"[14]

Zavitz and J.H. White identified rocky lands on the Canadian Shield as desolate wastelands where the stripping away of the soil and the loss of seed trees were preventing forest regeneration. They were able to have these areas reforested starting in the 1950s, when conservation authorities were being established in areas such as the watershed of the Moira River. The effort got another boost with the election of Prime Minister John Diefenbaker, a life-long friend of the visionary forester Richard St. Barbe Baker. His government created the Agricultural Rehabiltation and Development Act, which helped provide funding for conservation authorities and county forests to acquire wastelands for reforestation on the Canadian Shield. This photo, taken in 1963, shows one such reforestation effort: Red Pine planted in rocky sandy areas of Lanark County.

Stirred by the Rockland Forest, Porter's "militant" supporters soon spread the word of reforestation throughout Eastern Ontario. Lanark County and the United Counties of Leeds and Grenville signed on for the Agreement Forest Program. Lanark purchased 446 hectares of blow-sand desert in Lavant Township and reforested it, planting Red and White Pines as well as spruce. Eventually, the forest program in Lanark expanded to 4,451 hectares on forty-two properties. Today, forests now provide heron rookeries, deer yards, and nesting areas for species such as the Red-Shouldered Hawk that need large tracts of forests for survival. Leeds and Grenville also promptly acquired a 450-hectare tract of blow-sand desert in the headwaters of the South Nation River, today known as the Limerick Forest. The county forest system now involves 5,700 hectares on several sites. As before, these expansive woods provide habitat for wildlife needing large areas of forest habitat, such as the Wood Duck and Hooded Merganser; they also provide deer yards and nesting areas for raptors like the Goshawk.[15] In this manner, the ongoing campaigns of OCRA gave a boost to the protection of the Niagara Escarpment. One outcome was the creation of the Grey County Forest, which today involves 3,375 hectares in thirty-nine sites, many of which are now traversed by the Bruce Trail.[16]

The work of the OCRA was helped by small numbers of dedicated naturalists, fishers, and hunters. One of them, Monroe Landon, was a founding member of what remains Ontario's strongest environmental group: the Federation of Ontario Naturalists, now named Ontario Nature. Among its early efforts was the documentation of how the streams of King Township, north of Toronto, could be brought back to life through reforestation. What gave OCRA's campaign for conservation authorities a sudden inspirational jolt was the participation of a very influential group of citizens, the veterans of the First World War. The veterans, members of a Canadian chapter of Men of the Trees, were intrigued by the notion that armies could become reforestation corps. The charity's founder, British forester Richard St. Barbe Baker, had come to Ontario in the 1930s to participate in tree plantings with Zavitz and his assistant, A.H. Richardson. Richardson was a Boy Scout leader, and he used his connections to involve Boy Scouts in plantings at Scout camps near the Ontario Seed Tree Plant in Angus.

Baker's campaigns attracted the support of a remarkable elderly gentleman, Sir William Mulock, then the recently retired chief justice of Ontario. As chancellor of the University of Toronto, he would have been aware of John C.W. Irwin through Irwin's service on the university's senate and would have been aware of the forestry controversy stemming from Irwin's remarks. A staunch conservationist, it was Mulock's own interest in forestry that prompted him to create the Canadian chapter of Men of the Trees, one of his initiatives to get more people behind Zavitz's 1937 campaign for conservation authorities. He participated with Zavitz in the planting of an arboretum, which was designed to be an ecologically functioning forest around the new David Dunlap Observatory that had been established in 1935 in Richmond Hill. The superintendent of the project thanked Zavitz for supplying over 8,000 trees of different types on short notice. Mulock took part in ceremonial plantings at the observatory (the actual planting took place from May 8 to May 10) and Zavitz was told, "We hope ultimately to have something of which we both may be proud."[17]

The magnificent forest around the observatory remains a proud example of ecological restoration, but because it is currently under the threat of development, it is also a symbol of what can befall forests in areas of urban sprawl. Today, the 76-hectare forest is surrounded by the built-up area of Richmond Hill, except for a rail corridor that links wildlife to various natural areas possessing forest cover. The forest, a green oasis, has a herd of about thirty deer kept in check by a pack of coyotes. The forest also provides a refuge for a variety of birds, including the Pine Grosbeak, Brown Thrasher, White-winged Crossbill, and most spectacularly, the Barred Owl.[18]

A Liberal Party veteran — he had been the first minister of labour in Sir Wilfrid Laurier's government — Mulock was dedicated to high-minded democratic reforms. (He persuaded William Lyon Mackenzie King to enter public service as his deputy minister of labour.) Through Men of the Trees, he encouraged army veterans to get behind Zavitz's campaign for conservation authorities, and he orchestrated veterans' gatherings as campaigns for reforestation. One of the most important gatherings took place throughout downtown Toronto, involving great long parades over three days in August 1938. Over 60,000 veterans and 250,000 spectators were involved. The climax of the event took place on August 1, 1938, at a tree planting at Coronation Park near the Princes' Gates. Mulock had planned the event to carefully reflect the ideals of Imperial forestry, and this can still be seen today. At the centre of the memorial grove is a Royal Oak representing the King. It is surrounded by trees representing the Commonwealth Dominions, India, and Crown colonies. Circling this are a set of maples, each tree representing a unit in the Canadian Corps that fought in the First World War.

University of Toronto chancellor Sir William Mulock, who served as Canada's first minister of labour under Sir Wilfrid Laurier, played an important role in Zavitz's campaign to create conservation authorities in Ontario. This photograph, taken at Eaton Hall by the Globe and Mail *on June 4, 1940, shows Mulock (with white beard) surrounded by his family, planting a tree with the Canadian chapter of Men of the Trees. Enormous pressure was put on the Ontario government to provide legislation for watershed-based conservation authorities through similar spectacular events that involved thousands of First World War army veterans.*

Each tree was planted by a member of Men of the Trees, who was also an army veteran. Plaques to honour units of the Canadian Corps were unveiled after Mulock gave the signal. The veterans then gave an oath to uphold democracy, following a speech denouncing dictatorship, given by Mulock. The ceremony was concluded with a blessing from the Anglican priest Archdeacon F.G. Scott, who said that the Corps ideals, like the trees they planted, would soar "upwards towards heaven."[19]

In the spring of 1939, Mulock organized another spectacle during the royal visit of King George VI and his consort Queen Elizabeth. For this occasion, 125 hard maples were planted in Toronto from Fleet Street to the Old Fort York gate. A Salvation Army band played at the approach of the royal car as veterans in Men of the Trees held young maples to be planted by school children.[20]

Encouraged by King George's participation in this event, many Ontario Liberal legislators became involved in OCRA Field Days, the most prominent being the future premier, Gordon Conant, from the riding of Ontario County, now part of Durham Region. Since his riding was part of the Oak Ridges Moraine, he spoke from experience when he stressed the need for "the maintenance of streams and wells without which good agricultural land becomes barren."[21]

Enjoying the surge in his political support, Zavitz took some time off to travel to Prince Edward Island and give advice on reforestation. The invitation came from a Liberal cabinet minister, Walter Jones, a former student of his, who in 1943 would become premier of the province. The call was related to his government's efforts to reforest the newly developing Prince Edward Island National Park. Zavitz predicted that the pine seedlings he saw planted would achieve "remarkable growth." This he attributed to the appropriate "Maritime climate." During the trip, he and Jones developed strategies on how to stimulate reforestation throughout the island. On a return trip in the early 1960s, he noted, "Red and white pine are seen everyplace, in fence lines and abandoned fields, with no problem of reproduction."[22]

The trip to Prince Edward Island was also a delayed honeymoon for Edmund and his new wife, Margaret Irene (Henderson) Madden, the widow of J.F.S. Madden; Zavitz had remarried a year earlier, two years after the death of his first wife, Jessie. According to Kathleen Mackenzie, she also brought adult children to the marriage. The two families blended well, and one of Margaret's granddaughters became one of Kathleen's best friends. The couple drove to the island province by way of the Adirondacks of New York State, where they examined forest plantings by Bernard Fernow. Prior to coming to Canada, Fernow had briefly been a forester at Cornell University. While there, he had been responsible for the planting of the forest. A great controversy arose because Fernow's reforestation plans had included cutting down some trees as well planting new ones, his goal being to eventually have a healthier and more diverse forest. Opponents of Fernow hired a botanist, who described Fernow's work as inappropriate. Wealthy people owning cottages near the proposed cutting area protested to Cornell University, and as a result, Fernow's Department of Forestry was shut down. Now, three decades later, Zavitz visited the site to determine the outcome of Fernow's reforestation project. From the impressive state of the forest he saw, Zavitz concluded that Fernow had done a good job and that his critics had been wrong. These trees, despite the enormous outcry they once generated, were now a healthy forest.[23]

In 1939 Zavitz and John Irwin got an unexpected break that would allow them to confront Hepburn. The premier agreed to a motion by the Conservative leader of the opposition, George Drew, to hold an inquiry into the Department of Lands and Forests. Since the Liberals had a comfortable majority in the legislature, they could have easily blocked this motion. They allowed it to pass, however, believing that the fact there was a surplus in the budget of the department showed they were doing a good job — despite the fact that they were the last public authority in North America to have such a narrow view of forestry. In contrast, the American federal and state governments of the time were subsidizing their reforestation projects with huge amounts of money.

Hepburn's difficulties were compounded by his decision to attempt to drag out and delay the inquiry hearings, causing them to begin after

Canada and the other British Commonwealth states were embroiled in the Second World War. The considerable effort that Zavitz had poured into making conservation appear as patriotic in these circumstances, to avoid waste in wartime, would shake government policies. With wood so important to many wartime products, including the mosquito planes that proved vital to the Battle of Britain, waste was seen as intolerable. In his presentation to the parliamentary committee in 1940, Zavitz avoided confrontation but managed to make important points that would be elaborated with more political impact by Irwin, who was outside the government circle. In addition to describing the devastation caused by sulphur pollution from metal refineries, Zavitz decried the continued indiscriminate opening up of northern forests to agriculture. He stressed that "the policy of segregating and surveying forest lanes is a very important one" and urged that a study be carried out by agronomists and foresters. He indicated that it was impossible to determine if northern forests were being managed on a sustained yield basis, since his work was focused on Southern Ontario.[24]

Zavitz was also frank on the inadequate nature of reforestation efforts in Southern Ontario. He told the committee, "Our staff is entirely inadequate [in size] to carry out this work on the scale that we hoped to be able to carry. That is a question of money and a question of policy. I feel there are a lot of problems we have that cannot be solved by the men working at head office, there is too much traffic [deluged by phone calls]."[25]

Irwin followed Zavitz and effectively built on his testimony. Citing Zavitz's testimony on how his work was confined to Southern Ontario, he called him "the Chief Forester in name only." In response to angry questions from Liberal legislators, Irwin responded, "Mr. Zavitz is called the Chief Forester. That is his title, but his work is confined, as came out in the evidence before, entirely to Southern Ontario — practically entirely." He also amplified Zavitz's comments on the inadequacy of reforestation in the south, stating, "The present staff is inadequate even to secure the best utilization of the seedlings now available for distribution."[26]

The firings of fifteen foresters by the Liberals was denounced by Irwin, and he praised Quebec's combination of fire ranger and conservation officer roles to protect wildlife. Irwin challenged the Liberals' policy of refusing to invest public funds in reforestation, based on a notion that the budget of the Department of Lands and Forests collected from fees should cover its expenses. He warned that such attitudes would lead to the demise of the commercial logging industry in Ontario as the "vast amount of second growth ... being carefully nurtured in the United States" came to maturity.[27]

In response to Irwin, the minister of lands and forests, Peter Heenan, fell into a carefully laid trap. He defended the firings of foresters carried out by Noad, claiming that they could be replaced more competently by his "lumberjacks ... who learn in the bush." For Heenan these lumberjacks were "[the] most efficient fire fighters," since they "[knew] how to get around in the bush."[28] Heenan's outing of his anti-intellectual prejudices against schooling and foresters created a field day for the Opposition and the press. These foibles were highlighted in the Opposition's minority report. It chastised him for viewing forest conservation, reforestation, and research as minor matters and for making decisions "with little or no information before him from the well-trained experts in the Department."[29]

The effective use of the minority report to ridicule his government in the press caused Hepburn to change course. He shuffled Heenan from Lands and Forests to the Labour Ministry, while the deputy Walter Cain, who had clashed with Zavitz from the day of his appointment by Bowman, retired. Heenan was replaced by Norman Hipel, and Cain with Zavitz's long-time protege, Frank MacDougall. Norman Hipel, an MPP from Waterloo County, was the former mayor of Preston and quite familiar with the danger of floods on the Grand River, caused by deforestation. With amazing speed, the new team carried out reforms that Zavitz had long urged. Free grants of land for agriculture in the north ended and Crown land could no longer be purchased on installments. Crown land agents were professionally upgraded and placed under the supervision of regional foresters, all of whom were trained in forestry or another biological science. MacDougall had Zavitz visit the Quebec Ranger School, located near Quebec City. Established after the Second World War, this school provided a model for the combined training of forest rangers and conservation officers. The program was opened by the Department of Lands and Forest at Dorset, Ontario, and the school was located next to a complementary University of Toronto research forest.[30] This form of in-house training began to be discontinued in the 1960s with the growth of community colleges and new degree programs in conservation-related fields.

Zavitz's success in establishing conservation authorities became known on April 25, 1941, at a conservation conference in Guelph. The key figures who had fought alongside of Zavitz were brought together: Richardson, Porter, Landon, and representative of Men of the Trees C.R. Purcell. Also present were hunters and fishers, representing Ontario Federation of Anglers and Hunters; and professional foresters, representing the Canadian Institute of Professional Foresters and the Canadian Institute of Forest Engineers.[31]

The conference was helped by a spirit of co-operation between the wartime Liberal provincial and federal governments, now involved in post-war reconstruction. This revived, at least temporarily, the approach taken on long-term planning by the old federal-provincial Commission of Conservation. At the federal level, an advisory committee on post-war reconstruction was established; the research director was Leonard Marsh. A few years earlier, he had been a leading force in the Co-operative Commonwealth Federation (CCF) "brain trust" — the League for Social Reconstruction. A committee from the Guelph conservation conference met with the federal reconstruction body in August 1941, when it was agreed that a pilot study of a watershed would be undertaken. This was to be the model for the new conservation authorities as fostered by provincial legislation. It was hoped that this new initiative would achieve what the Commission of Conservation had been trying to foster, but unable to achieve.

Zavitz, in the spring of 1941, found himself in an excellent position to realize the Commission of Conservation's unmet dreams. He had seen his friend James White's plans for the massive Trent canal watershed fail because the Ontario government would not support recommendations. This example encouraged him to establish a pilot reforestation project on a smaller watershed, and one, moreover, whose problems he had understood from childhood — the Ganaraska. He had planted his first forest here in 1905 and had kept in touch with his maternal relations, the Squairs, who still owned his grandfather's farm. The task of writing the pilot Ganaraska Watershed Survey was assigned to Zavitz's long-time assistant, A.H. Richardson. In the summer of 1941, high-school science teachers from across Ontario were assembled to assist him. Their presence there was helped by the extension of the school holiday period to October 1, a move intended to help in the war effort by having more high-school kids working on farms. The study was also helped by an OCRA Field Day, when participants toured the reforested Durham County Forest, examined a cut-over in a stream valley, viewed desert-like conditions, and a rare example of an old growth forest.[32]

Richardson's report thoroughly documented the disasters on the Ganaraska arising from deforestation. It expressed dismay at how the stream originated in "springs at the base of sand dunes" created from deforestation, whose tributaries "wore deep gullies" into the Oak Ridges Moraine. He used an 1878 atlas of Northumberland and Durham Counties to show how headwater streams then had extended "much farther up the moranic slope, which at that time was well wooded." Richardson found that many of "these dried-up watercourses" could still be followed.[33]

Richardson's study contained two key recommendations: there should be massive reforestation and existing and future forests should be protected by a new regulatory tool — the tree-cutting bylaw. The report called for "20,000 acres" (8,093 hectares) of the Ganaraska's headwaters on the Oak Ridges Moraine to be reforested. This would be carried out under the direction and ownership of a future conservation authority, to be created through provincial legislation.[34]

Apart from ownership by the conservation authority, forests were to be protected through tree bylaws that also included private-land owners. The intention here was to prohibit clear-cutting of forests, which removed the tree canopy. To establish a benchmark, Richardson recommended "cutting to a diameter limit, that is all trees below a certain diameter — for examples, five inches — should not be cut." By outlawing the cutting of smaller, younger trees, such tree bylaws would ensure that the forest ecosystem would survive the impact of selective logging. Richardson's "A Report on the Ganaraska Watershed" was released in the spring of 1944. By 1946, the Conservative government of George Drew had implemented its two key recommendations: conservation authorities were established to apply the Agreement Forest Program and engage in massive reforestation, and the passage of the Trees Act in 1946 permitted municipalities to outlaw clear-cutting on private lands.[35] The fact that reforestation on the Ganaraska began so promptly following the passage of both the 1946 Conservation Authority Act and Trees Act allowed Richardson's goals for ecological restoration and watershed health to move ahead.

In other parts of the province, delays over the creation of authorities (which required complex municipal approvals) caused land prices to inflate, and later, program dollars to dry up. This made for less dramatic results. What was achieved on the Ganaraska, however, was impressive, with 43 percent of the watershed now in forest cover. Deserts, drought, and floods have vanished. In addition to the reforested headwaters, good riparian cover now exists on the Ganaraska from its Oak Ridges Moraine headwaters to Lake Ontario, cooling the stream for salmon and trout. Brook Trout have increased in both numbers and size, and trout from the Ganaraska are now used to restock depleted populations in other parts of the province. Deer and beaver populations, virtually gone at the time of the 1943 survey, have also recovered.

While Zavitz is recognized as the "father of reforestation in Ontario," his parenting of tree-cutting bylaws is less celebrated. His experience with the Agreement Forest Program in acquiring county forests before the Second World War, however, had taught him the need for the creation of such bylaws. He was shocked to learn that private landowners, after being paid for the land by municipalities, would be allowed to "cut out timber before giving title." In a conference held in London in 1944, Zavitz detailed the need for laws to protect forests on private lands:

> It has been demonstrated that forests can be put back on the most barren lands. What is required today is an organized effort to rehabilitate these areas. This means trained men and money, with legal authority over the area in question to prevent private interests from again creating barrens and destroying the natural forest protection of vital watersheds. Considerable public sentiment has been aroused against the wholesale destruction of private woodlands. Legislation, and machinery to administer the enforcement of forest protection on our important watersheds should at least be given first consideration.[35]

Zavitz's London remarks were part of the address "Reforestation as a Means of Controlling Run-off," which was intended to balance the perspectives of others who relied more upon dams and other constructed works to control floods. His remarks should be seen as quite prophetic, since it was not until the 1970s that such engineering approaches to flood control began to fade away; this change in ideology reflected sterner requirements for justification of such approaches. Zavitz stressed that the potential for forests to moderate run-off was disguised by the pitiful state of many of the Ontario woodlands that survived. This was because "many so-called woodlots [did] not fulfill the requirements of natural forests." They were "composed of scattered, inferior or defective trees ... the ground cover [was] imperious soil, with grass and weeds, giving very little protection to the absorption of rainfall and runoff."[36]

Zavitz explained that healthy forests would do a better job in protecting watersheds than the dying ones that haunted the barren landscape:

Rain falling on the forest floor does not run off suddenly, but is taken up by the soil and stored for future use. Snows in winter are held under the forest cover and allowed to melt slowly, the water gradually reaching [deeper] levels of the soil. It has been found that frost does not go down as deep in the forest as upon the open fields. The fact that forest cover prevents rapid run-off of rains and melting snow makes it important that a good percentage of the watersheds of our rivers be kept under forest cover. I assure you that the consensus of opinion is that forest cover tends to equalize the flow throughout the year by making the low flow stages higher, and the high stages lower.[37]

At the London Conference Zavitz laid out a vision for watershed protection based on the reforestation of the Niagara Escarpment and Oak Ridges Moraine, now enshrined in the land-use planning controls of the Ontario Greenbelt. He told participants, "Look at the area where the headwaters of the Grand, Saugeen, Nottawasaga, Humber and Credit originate, then travel across the Iroquois and Beach Ridges [now Oak Ridges Moraine] from the source of the Don to the Ganaraska ... I can assure that in these are many thousands of acres, which never should have been cleared but which should have been kept in forest, for forest crops and watershed protection."[38]

In 1945 Zavitz authorized one of his foresters, G.H.U. "Terk" Bayly (in 1966, Bayly would become deputy minister of the Department of Land and Forests), to undertake a census of all the forests created under the co-operative planting program, with the intention of providing more evidence of the need for tree-control bylaws and improved extension services. He found much room for improvement. The best results came from plantations of tough coniferous species planted on farmlands across Ontario. The worst damage to planted trees came from livestock grazing on seedlings.[39]

Zavitz made presentations to two government investigations on forest conditions between 1946 and 1949. One presentation involved the Royal Commission on Timber. Its report, delivered by the chair of the commission, Major General Howard Kennedy, was released in 1947 and was followed by the Ontario Legislature's Select Committee on Conservation, published in 1950. Although these committees' findings were released after the forestry reforms were legislated, the publicity they generated helped greatly to secure more funding and generate public interest, even inspiring people like Mel Swart to become conservation activists.[40]

Zavitz told Major General Kennedy that his experience with the co-operative planting program showed the need for more "trained men in the field" to assist farmers. Tree bylaws were needed to stop lumbermen "who [bought] the timber only to strip it." He urged that the Fire Districts established under the Forest Fires Prevention Act be extended to cover the whole province. Regarding the Oak Ridges Moraine, Zavitz testified, "The conditions at the headwaters of the Ganaraska River have to be seen to be believed."[41]

Edmund's cousin, Harold, whom Zavitz had appointed as zone forester in the extreme southwestern part of Ontario where resistance to reforestation was the strongest, explained the challenges he had in increasing forest cover to Major General Kennedy. He found that drainage schemes in Kent and Essex Counties caused trees to die from "radical changes in the water table," causing municipalities he had tried to get involved in reforestation to lose interest. Harold also testified that Edmund's 1908 wasteland study was still the most detailed report of spreading deserts in Huron and Lambton County, both of which continued to be damaged by forest fires.[42] Monroe Landon also made detailed submissions to Kennedy and the Select Committee. He spoke about the damage to forests in Norfolk County by forest fires, which he believed were being set to clear the forests before the tree bylaw protections could be implemented. He urged that the fire marshal of Ontario investigate the blazes and that any landowner convicted be required to reforest at his own expense.[43]

The Ontario Federation of Anglers and Hunters strongly endorsed Zavitz's reforestation plans. They told Kennedy, "Dependence of the land determines the very occurrence of wild animals. In far too many cases where the land has been completely cleared for agricultural purposes, the situation was very grave indeed. Our streams have disappeared, soil erosion is progressing and without ample stores of water and soil there can be no sustained plant or wildlife."[44]

Stressing the need to restore and protect forests, the outdoorsmen mocked other efforts to encourage game such as artificial stocking and predator control. They reported that forest destruction had made 1946 the worst year for deer hunting in Ontario. Zavitz's call for the merger of the Game and Fish Department with Lands and Forests was strongly endorsed. They also backed his testimony that "trained men should be out and on the land to work right on the field and woodlands along with the farmer."[45]

By 1947 the basic shape of Ontario's conservationist reforms was in place. Edmund Zavitz would spend the last two decades of his life happily implementing his vision of ecological restoration. Essentially, he had created all the laws, policy, and government funding to protect and expand forests, both in the north and the south of the province. Crown lands were now under the control of foresters, and farmers were being kept out of marginal lands in the north. Placing Crown lands under the control of foresters would lead to important reforms in land-use planning, shoreline reservations, and reforestation. In the south, conservation authorities would provide leadership to dramatically expand forest cover.

Zavitz's delight in securing the passage of his conservationist "new deal" was enhanced by the joy of the return of his son Ross from five years of artillery service in the Second World War. He had served with distinction in the dangerous Italian campaign and achieved the rank of major. Ross returned to the Forestville farm, originally purchased by his father as a hobby activity, which ultimately had turned into a full-time venture. When he returned, Edmund was so proud of his son that at the welcome home celebration he donned Ross's officer jacket and hat.

— NINE —

Implementing the Vision

During the last twenty-one years of his life, from 1947 to 1968, Edmund Zavitz would delight in implementing the vision of an ecologically restored forest landscape. It was now possible through the legislative statutes, funding mechanisms, and government programs he had created. The natural world and its dependents within the whole of the province would benefit from his vision. At long last, Ontario would be brought to the standards of care for the earth that had been accomplished earlier throughout the Commonwealth and the United States.

By 1947 he had brought the necessary programs, laws, staff, and funding into being, but the scale of the task might have daunted a lesser man. Since 1904 he had passionately dedicated himself to public service in order to reforest Southern Ontario, but its percentage of land in forest cover had actually decreased. The figure for 1943, 9.7 percent forested land, was the lowest in recorded history. The significant changes he would bring, however, can be seen in the remarkable expansion in the last two decades of his life. By 1963 he had achieved a tripling of Southern Ontario's forest cover, which then stood at 25.2 percent.[1]

A symbol of the green tide advancing against the still-marching sands took place on May 14, 1947, with the planting of the first tree in the Ganaraska Forest. A photo caption marking this occasion recorded the significance of this event, organized by the Ganaraska Conservation

Authority. Noting that the "Honourable Dana Porter, Minister of Planning and Development, when planting a tree in the Ganaraska Forest, threw off his coat and vest," it appeared "symbolic" of what is to take place on the Ganaraska and other Ontario watersheds.[2]

It was Zavitz's zone foresters who provided the skill and training needed for conservation authorities to purchase, reforest, and manage the land. Yet, despite the backup support, municipal councils were reluctant to use their new clout to create conservation authorities for their areas. Zavitz's friend Watson Porter attempted to form a single comprehensive Thames River Conservation Authority. However, communities downstream were less vulnerable to floods than upstream London, and he was forced to settle for the creation of an Upper Thames authority only.

Flooding problems resulting from deforestation were the main reasons behind the creation of most authorities during Zavitz's lifetime. The first two were created for Etobicoke Creek (now part of the Toronto and Region Conservation Authority) and for Ausable River, which empties into Lake Huron near Sauble Beach. In both cases, the initial reticence of municipal councils to plant forests was overcome by severe flooding and the resulting lawsuits.[3]

The Ganaraska Authority, which Zavitz had done so much to create, was the third in Ontario. Despite the massive flooding the communities faced, with downtown Port Hope the only major town in the watershed having to be evacuated annually, municipal councillors protested that they did not want to spend money on reforestation. This reluctance was finally overcome when they decided to limit reforestation expenses to $5,000 annually for the first five years. Once work got underway, however, action was swift as Dana Porter's vest-stripping was intended to symbolize. By 1991 the Ganaraska Forest created by the authority covered 4,046 hectares. For many years, much of what the authority did was to organize Boy Scout tree planting days and provide public tours of the new forests. In addition, the Ontario County Forest covered another 468 hectares and some 390 hectares of private lands were reforested. This meant that 43.6 percent of the watershed was in forest cover. Eventually, these and other private forests were successfully kept

— *Implementing the Vision* —

Courtesy of Ed Borczon.

The goal of the Ganaraska Conservation Authority, created in 1948, was to create a massive forest in the headwaters of the Ganaraska River on the Oak Ridges Moraine. This photo, taken around 1950, illustrates the difficult circumstances encountered as trees are planted into the scarred landscape of the moraine.

in forest cover by effective tree-bylaw regulations. Within ten years, the plantings had led to streams flowing more evenly and the reduction of spring flooding and summer low-flow conditions. In addition to Brook Trout thriving again, as Zavitz had predicted in his 1908 study, the deer, beaver and Wild Turkey had returned.[4]

In 1949 Zavitz retired from the position of director of reforestation to become a consultant to the Department of Lands and Forests for five years. During this time, he was engaged in reforesting the headwaters for streams that now form the basis of the Ontario Greenbelt, essentially the Oak Ridges Moraine and the Niagara Escarpment. That same year, 809 hectares of the Ganaraska Forest had been planted on the Oak Ridges Moraine, and the zone foresters working under his supervision acquired 1,618 hectares for the Bruce County Forest on the Niagara Escarpment. Funding formulas were devised to give bonuses for the purchase of

Ted Jenkins, photographer for the Department of Lands and Forests, took this picture near Collingwood in 1954, showing major reforestation underway on the Niagara Escarpment. Protection of watersheds through reforestation and the acquisition of public lands provided the basis for Ontario's Greenbelt, created in 2005.

existing forests, reflecting the fact that these areas were under threat and had lower maintenance costs than reforestation projects.⁵

In 1947 Zavitz's publicity efforts to promote the virtues of reforestation in Southern Ontario included the involvement of old friend, Gordon Dallyn, then the editor of *Canadian Geographic*. Zavitz had appointed him district forester for Tweed in the 1920s. Together, they produced the article "Fifty Years of Reforestation in Ontario" for publication in *Canadian Geographic*. The article was later republished as a booklet by the Department of Lands and Forests, and this version included some fifty-three photographs, all of which had been taken by Zavitz. His photographs showed the stark gullies and sand dunes along with a few solitary White Pines in the headwaters of the Ganaraska. Below, a caption warned, "A river started here." Such scenes of devastation were coupled with hopeful images of reforestation. One showed four Boy Scouts planting the Angus Municipal Forest in Simcoe County during the Great Depression, then returning to the scene as adults, happily visiting the wonders they had nurtured.⁶

Another enjoyable task for Zavitz was supervising the erection of a memorial cairn in honour of J.H. White at what was then Forestry Station No. 2 at Turkey Point. Dedicated after White's retirement from the Faculty of Forestry at the University of Toronto, the memorial thanked him for his "inspiration and guidance to his many students in those early years."⁷ The ceremony took place on September 12, 1949, with Zavitz unveiling the cairn in the presence of thirty-two members of the Forest Engineers Society, many members of the general public, as well as folks from the St. Williams Station, where the day was declared a holiday. Forestry Station No. 2 was renamed as the J.H. White Forest, a name that lasted only up to the creation of Turkey Point Provincial Park in 1958.⁸

Shortly after this event, another major step took place in Halton Region, where a growing system of county forests protected a major stretch of the Niagara Escarpment. In 1949 Halton passed a bylaw to aid farmers in fencing their woodlots to protect them from roaming livestock. Halton provided a grant of money from the county budget equal to the prevailing cost of installing fence wire with a single barb to enclose a field. This was the first municipal recognition of the costly problem of fencing, something that had long plagued municipal councils in Ontario, Zavitz, and even his predecessors such as Judson Clark.

An more dramatic event also took place in Halton County — the reintroduction of the beaver. Zone forester, I.C Marritt, after achieving permission from neighbouring landowners, took the audacious step of bringing beaver back to the headwaters of Oakville Creek on the Niagara Escarpment. For some time, he had been facing the problem of Oakville Creek being too warm in the summer and too low for its threatened Brook Trout population to survive. Beaver dams would help pool the water during this critical summer period for the Brook Trout, thus reducing the temperature of the water. Marritt recorded the day, October 8, 1949, of the first beaver dam to be constructed near the City of Toronto in over a century.[9] The return of the beaver to the predominately agricultural areas of Southern Ontario was another feature of Zavitz's restoration of its forests, with an eye to building for the future. But he made an equally formidable contribution to the protection of northern forests.

In Southern Ontario, Zavitz's conservationist forestry practices were initially applied to rescuing forests from encroaching sand deserts, and then later applied to transforming northern forests from largely barren hectares of rock into once-more flourishing stands of trees. By 1928 he had essentially rescued the northern forests and was on the brink of putting Crown lands under the supervision of foresters. The big step in this direction had been the creation of provincial forests, replacing the former category of forest reserve. Zavitz had planned to modify logging practices so they would, as in the forests of Algonquin Park, be compatible with recreational pursuits such as canoeing, fishing, and nature observation. His work in this regard was disrupted when Hepburn's government fired massive numbers of both foresters and game wardens, as noted earlier.

When Zavitz ultimately triumphed over the narrowed-minded logging interests championed by Deputy Minister Noad in 1941, the reforms he had fought for came through in gradual stages. Surprisingly, even with Zavitz's long-time positive reputation in the field of forestry, there was considerable resistance to embracing even modest reforms in the north. It took two years after the reorganization in 1943 to close the northern Kirkwood District to agriculture, even though it was known that attempts to farm its sandy soils were one of the reasons for desert conditions there. Likewise, even with the combined forest and farming soil studies Zavitz called for in his 1939 testimony to the committee of the inquiry into the Department of Lands and Forests, the response to the need for reforms was slow. The most extensive of these studies, the Glackmeyer Report of 1960, was needed to finally curb the spread of agriculture in the northern Clay Belt. This was coupled with the successful zoning of Crown lands to prevent their sale for agriculture.[10]

Although farming was limited in the north, in places like Kirkwood it created environmental problems such as flooding and deserts, as had happened earlier in the south. In dealing with these problems, Zavitz had the support of at least one conservationist-minded member of the paper industry. His friend Mel Swart had the support of his employer, Ontario Paper, which was located in Thorold but obtained its wood supplies from the north. Farming in the Clay Belt was becoming disruptive to the long-term forestry supply for paper makers. As pointed out in the Glackmeyer Report, farmers in the Kapuskasing area would abandon their farms after stripping their properties of trees for sale as pulpwood. Under such circumstances, the trees needed for the paper industry would not regenerate.[11]

The problem of local deforestation caused by inappropriate agricultural practices was a key reason for the launching, in 1946, of the first northern Reforestation Station in Fort William under Zavitz's supervision as director of reforestation. This station, like those in the south, was surrounded by a block of a 202-hectare forest grown from seedlings to foster more local watershed forest cover. What had put the problem of deforestation in the Fort William area on the radar screen was a severe flood in the communities of Fort William and Port Arthur in 1941. When this flooding began to reappear annually, the local response was to create the first northern conservation authority.

The Fort William Reforestation Station helped with the tree-planting efforts of the Lakehead Region Conservation Authority, which became the only northern entity to use Zavitz's Agreement Forest Program. The authority acquired the Mills Block Conservation Area, a swamp forest of Black Spruce and cedar north of Thunder Bay, now designated a provincially significant wetland. One of its purposes is public education in beaver-dam making and local ecology. To date, the authority has planted 450,000 trees and has reforested 2,500 hectares. To further encourage reforestation of marginal farmland, the conservation authority has a Private Landowners Tree Seedlings Assistance Program.[12]

Apart from maintaining their own forest holdings, today's conservation authorities in the north are concerned with protection measures in private lands, such as keeping development away from flood-prone areas and having culverts properly constructed so as to not disrupt fish habitat. They are not managers of the vast Crown lands of the north. Zavitz's influence with these northern Crown lands would prove to be critical in his last years.

Paradoxically, the symbol of Zavitz's impact on the management of northern Crown lands is the disappearance of the designation of provincial forest in 1960, which he had created in 1928. The use restrictions on provincial forests — prohibition of sale for farming combined with various environmental controls on logging — had, by 1960, become the norm throughout the Crown lands of Ontario that were being harvested commercially. This harvesting was based on timber management agreements and leases, a system that goes back to 1848, with the government retaining the land and the logging companies getting the trees.

During the reign of Hepburn's government, the provincial forest designation had lost all meaning, but the reorganization of the Department of Lands and Forests in 1941 changed this situation. As former forest reserves where logging was more restricted than other Crown lands, many of these areas eventually became provincial parks. In 1943 Sibley (now known as Sleeping Giant Provincial Park) was the first provincial park to be created from a provincial forest. After the reorganization of the Department of Lands and Forests, foresters soon discovered that it was a major effort to curb poaching — a result of the chaos of the Hepburn years. However, efforts to protect wildlife persisted, and today, moose, bear, and even a rare Woodland Caribou herd have made a comeback.

For the first time, district foresters working on Crown lands in the north under licence agreements had a new power, the power to protect the habitat of the Brook Trout, which is an excellent indicator of ecosystem health. Zavitz had used the Brook Trout as his indicator in Southern Ontario, and in his wasteland report of 1908 and subsequent conservation studies, he encouraged more reforestation of the watershed and riparian cover to shade streams. The Brook Trout is an excellent indicator, since it requires cool, shaded streams that are free from polluting sediment to survive. It also sends out warning signals, such as not growing to its normal size, before disappearing entirely. Shoreline reservations were a belt at least ninety-one metres wide, designated to protect trout habitat on those Crown lands used for logging. These shoreline strips also had fire-control benefits since they provided a break in "vast areas… of dangerous [logging slash] which in a fire is most difficult to control."[13]

During Zavitz's period as a consultant to the Department of Lands and Forests, these shoreline reservations for Brook Trout were established in the Spruce Falls licence area around Kapuskasing. The professional foresters employed by the Spruce Falls Company (then owned by New York Times, now by Tembec) co-operated with foresters of the Department of Lands and Forests in protecting Brook Trout habitat. They did this through buffers that cooled streams and by keeping sediments out of streams around Kapuskasing. The company also provided similar reforestation projects in the north by having people hand-plant seedlings in the openings of clear-cut strips.

In 1949 Zavitz was given a retirement banquet and a fine radio for his years of service to the Ontario government. During his remaining five years as a consultant with the Department of Lands and Forests, one of his most pleasant experiences was to implement his reforestation agenda in Eastern Ontario. He also saw the White Pine thrive once more as a commercial species in Algonquin Park; this was a culmination of policies he had long pushed forward. He had called the White Pine "the tallest and most stately tree" in eastern North America. Its regeneration was the key goal of the Petawawa Management Unit, which was created in 1945 and named after the river flowing through the eastern section of Algonquin Park. Foresters working in the unit modified the natural regeneration of pine using the strip-cutting shelterwood system, so-named for the practice of having the older trees provide shelter for the younger trees, a mid-way situation between selective logging and clear cutting. Cutting, however, when it takes place (usually there are long intervals, sometimes twenty years, between cuttings), is more intense than selective cutting, since entire rows of trees are removed. Some 77 percent of White Pine in Ontario is harvested by this method.[15]

Zavitz had a very dedicated team to encourage reforestation in Eastern Ontario. Ferdinand Larose remained active till his death in 1954, and the newly opened Howard Ferguson Reforestation Station in Kemptville focused its activities there. The South Nation River Conservation Authority cuts through Eastern Ontario, almost to the St. Lawrence River. The Larose Forest, which Larose helped to found in 1947 when he used the Agreement Forest Program to reforest 2,332 hectares, is in its watershed. County forest systems initially launched by Larose and Zavitz during the Depression to reforest wastelands now acquired swamp forests for wetland protection.[16]

One useful tool in the package of reforms that Zavitz persuaded George Drew to undertake as part of their new agreement of 1946 was permitting townships to take part in the Agreement Forest Program. This proved especially important around Ottawa, since Carleton County delayed its participation until 1965. One of the Carleton townships, Charlottenburg, had moved ahead in 1954 but later transferred its agreement forest to the Raisin River Conservation Authority, which was created in 1963 to serve the western part of the area. A 365-hectare forest, created by Cumberland Township, later became an important moose passage area between the Larose Forest and the Blue Mer Bog in the Ottawa Greenbelt.

Today, the Township of Marlborough (Carleton County) is largely remembered for the forest it set in motion. With the help of Zavitz, it acquired ninety hectares under the Agreement Forest Program in 1953. Eventually, following Carleton County's decision to become a participant, the Marlborough Forest became much larger and protected 3,400 hectares. In 1970 the county was amalgamated into the Regional government of Ottawa-Carleton. The core forest area, held in public ownership, was complemented by adjacent provincial Crown lands and protected from urban sprawl on a surrounding 14,973 hectares of private lands by zoning bylaws.

The Marlborough Forest has become one of the most southerly extensions in Ontario of the Snowshoe Hare and its predator, the Canada Lynx. It is a refuge for a number of rare species, such as the Black Tern, Red-Shouldered Hawk, Pileated Woodpecker, and the Marsh Wren, and became a staging ground for the restoration of the River Otter to Eastern Ontario. For thirty-seven kilometres, the Rideau Trail from Kingston to Ottawa snakes through this forest, the largest stretch on the 387-kilometre trek.

The creation of the Marlborough Forest was another example of ecological restoration in areas that had been destroyed by fire. Most of the original forest had been burnt down to bare rock by massive and frequently repeated fires, generally set by farmers in their efforts to clear land in Carleton County. Such burns created alvar-like conditions, where the soil is so thin that it can only support low-lying bushes, although at one time the soil was deep enough to hold gigantic pines. Massive charred stumps of White Pines can still be found throughout the forest today, as can the ruins of stone fences that line the now-abandoned farms.[17]

One of Zavitz's most spectacular victories in Eastern Ontario was secured in 1960 when the National Capital Commission joined the Agreement Forest Program to reforest the Greenbelt around Ottawa. Under its provisions, the Department of Lands and Forests reforested and managed 804 hectares of Greenbelt land. One of its most significant forested tracts is the Blue Mer Bog in the eastern part of Ottawa's Greenbelt, close to Larose Forest. It became another part of the range for Ottawa area's moose herd. It also became an important breeding area for the Northern Hawk Owl and Sandhill Crane.[18]

Studies Zavitz had commissioned in the 1920s revealed the healthy forests of Algonquin Park surrounded by the wastelands of adjacent Renfrew County. In 1952, when he flew over the area from Ottawa to Algonquin's Opeongo Lake in a Beaver aircraft, this contrast became even more evident. He could take comfort, however, in knowing what was to come. Renfrew County had signed up for the Agreement Forest Program on December 24, 1951.[19]

The Renfrew County Agreement Forest, which now covers 6,474 hectares in fifty tracts, ended its sharp contrast with the eastern edge of Algonquin Park and helped expand the range of Algonquin's large mammals such as moose and bear. Thirty of its forest tracts protect Brook Trout habitat. One of the forest's purposes is to protect evidence of doomed farming ventures, such as the still-lingering building foundations. The ruins of these farms are reminders of the desperation of their former owners, who, during J.H. White's Trent survey research, pleaded with him to buy the land and release them from their dire poverty. He was not able to help them, since the government decided not to implement recommendations of his report, which would have seen their land purchased for reforestation.

Zavitz used the new conservation areas in those wastelands, already identified in his 1908 report, to realize White's plans for the reforestation of Canadian Shield areas outside Algonquin's borders. The task was taken on by the Moira Valley and Napanee Region Conservation Authorities, today merged with the Quinte Conservation Authority. Their planning recognized the need "to preserve and restore the forest cover of the northern portion of the watershed comprising of that region lying above the Laurentian Shield." The 1955 Moira watershed study contained photographs quite similar to those which White used in 1913 to document deforestation. The report laid out plans to reforest these barren wastelands by planting hardy native species such as

White Spruce, Sugar Maple, White Oak, and Bur Oak into crevices in the bare rock at the rate of "two hundred trees to the acre." Accordingly, plans were developed to restore the vast "semi-barrens" of Kaladar, Sheffield, and Madoc Townships; a component of this plan was to "gradually build up cover of duff [organic matter in soil] which it would take to retard runoff" over the existing barren rocks. In this way, the Agreement Forest Program restored some 9,937 acres [3,657 hectares] in the Moira watershed.[21]

At the height of his reforestation successes in the 1950s, Edmund Zavitz was given a number of important awards and distinctions. The first, in 1951, came from the Canadian Forestry Association. This was followed by two Doctor of Law degrees; the first came from McMaster in 1952, followed by another from the University of Toronto in 1957. What was especially appropriate in the University of Toronto ceremony was that in addition to a former student of Zavitz, H.R. MacMillan, yet another person who helped make Zavitz's successes possible was presented with an honorary doctorate. This was Joseph A. Bedard, the veteran architect of Quebec's forest service who had enjoyed the support of the powerful Quebec Catholic Church. He had pressured Queen's Park into supporting Zavitz at two critical points in the latter's career. The first was in 1912 when Zavitz gained the power to regulate railways. The second came in 1934, when a personal visit to Hepburn helped to wind down Frederick Noad's period in power.[22]

Zavitz's previous honours were topped by the Ontario Archaeological and Historic Sites Board memorial plaque at the St. Williams Forestry Station. It was unveiled on September 4, 1957, by the minister of lands and forests, the Honourable Clare E. Mapledorm. The plaque acknowledges Zavitz as "the father of reforestation in Ontario," recognizes that the St. Williams Station was founded on "100 acres of wind eroded sandy land," and praises his leadership in restoring "large areas of waste lands" to productivity. It is interesting to note that the wording of the text does not detail his struggles to protect the north, which came to fruition only after a series of bitter conflicts. A major error in the plaque's wording indicates that Zavitz was deputy minister of the Department of Lands and Forests, while he actually was deputy minister of forests from 1926 to 1934. Interestingly, the actual deputy minister of the Department of Lands and Forests was Walter Cain, who had clashed with Zavitz over Cain's support of farm and logging interests hostile to conservationist forestry. Could it be that the intense loyalty of the department staff resulted in the elevated office attributed to Zavitz? They would be only too painfully aware of the traumatic past.[23]

Edmund Zavitz and Clare Mapledorm, minister of the Department of lands and forests, unveil a plaque to commemorate the founding of the St. Williams Forestry Station in 1908. A plaque recognizing Zavitz, erected by the Ontario Archaeological and Historic Sites Board, was unveiled on September 4, 1957. The wording of the plaque tends to downplay the magnitude of Zavitz's struggles. Photo by Ted Jenkins.

The effort to honour foresters like Edmund J. Zavitz, Joseph A. Bedard, and Harvey H.R. MacMillan in the 1950s was the result of a shared appreciation of the positive ways that a conservationist approach to forestry had altered the landscape of the North American continent. Three other foresters were also honoured: Wallace A. Delaney of Toronto, Robert W. Lyons of Toronto, and John Allen Gibson of New Brunswick.

In 1953 Zavitz went to Ann Arbor, Michigan, for the forestry alumni gathering of the state university. While his mentor Filbert Roth and P.L. Lovejoy, a fellow student who had done so much to foster reforestation in Michigan, had died, friends from his student days were there to greet him. Of these the most notable was the recently retired chief forester of Michigan, Marcus Schaaf, who had replaced Roth as chief forester in 1910 so Roth could return to full-time teaching. While enormous areas of the state had been reforested under Schaaf's leadership from 1910 to 1949, greater stages in ecological restoration were to build on this foundation. This came in the 1980s when, with the help of Ontario's Ministry of Natural Resources, Michigan restored Pine Martin, Fisher, and moose populations.[24]

The next year, Zavitz made a journey to the former winter home of his mother and step-father as part of a trip to the American Gulf States. During the 1920s, he had stayed with them, needing a rest from his arduous work. In "Recollections, 1875–1964," Zavitz wrote, "At that time we saw a ground fire burning through the southern pine forests. Little attention was paid to control, and it was looked on as a benefit for grazing." While in the American south, Zavitz saw that "considerable reforestation was being done and frequently, while there were scattered seed trees, they carried out restocking as they could not wait for natural reproduction."[25]

It was not possible for Zavitz to create the same dramatic landscape transformation in south-central Ontario — with the return of large mammals such as moose and bear — that he was able to achieve in Eastern Ontario. This was largely the result of urban sprawl inflating the price of land and making it much more difficult for the government to pay for land to establish forests. Zavitz addressed the problem of developers buying land before his zone foresters could reforest in a March 13, 1951, letter to the York County Council. Zavitz warned that potential properties were being lost through the combination of long approval processes and escalating prices. He pointed out that spending limits of $50 an acre were below the market rate of $55 to $60 an acre.[26]

Zavitz warned the government of the need to act with haste to acquire forests in York County because of the work associated with the Don River watershed plan. In contrast to all the other watershed plans in the province being prepared by Richardson's Department of Planning and Development at this time, urban sprawl was the big threat to the Don's ecosystem. The watershed plan's call for the creation of a Don Forest in its headwaters on the Oak Ridges was not realized as municipalities were unwilling to pay, and the Don became the most polluted stream in Canada as 82 percent of its watershed was urbanized. Where the plan was successful was the realization of the need for a "greenbelt" in the Don's valleys, achieved in part because of the threat of flooding and the damage it could do to buildings.

When later, in 1954, Hurricane Hazel devastated valley land, acquisition sped up. The process, however, is still ongoing.[27] The delays demonstrate some of the conflicts stemming from the pressures developers put on Zavitz's conservation plans for an area so close to the rapidly expanding City of Toronto. Although the goals were the same — reforestation of the Oak Ridges Moraine and improvements in fish habitat and water quality — the results were less substantial than in the Ganaraska watershed to the east.

The watershed plan for the Central Lake Ontario Conservation Authority called for the reforestation of 4,300 hectares along the Oak Ridges Moraine. The authority, however, was only able to acquire and subsequently reforest 446 hectares in what became the Long Sault Conservation Area. Spurred on by the knowledge of the benefits of reforestation, provincial funding to the Oak Ridges Moraine Foundation over the past decade resulted in the purchase of the Crow's Pass, Rahamni, and Enniskillen Conservation areas on the Oak Ridges Moraine.[28]

Zavitz's major obstacle in southwestern Ontario was the high value of fertile agricultural land. The greatest challenge was in Essex County, where his ideas for creating a conservation authority were not taken up until after his death. Here, however, once established, the Essex County Conservation Authority had far better results than similar agencies on the American side, namely those of the Corn Belt area. While Essex has only 7.5 percent of its landscape in some form of natural habitat today, the situation in Iowa is only 1 percent.[29]

Two of Zavitz's old friends, Nelson Monteith and Watson Porter, played major roles in the formation of the Upper Thames Conservation Authority in 1947. Over time, the Upper Thames Authority almost doubled its forest cover, which grew from 6.7 percent to 11.5 percent. Although still well below the contemporary goal of 30 percent forest cover, the achievement helped avoid a repetition of the flooding of the Great Depression. During the last years of his life, Monteith was able to chair the Upper Thames Conservation Authority's land-acquisition committee. He played a major role in the acquisition and expansion of forests in the Ellice and Gads Hill Swamps, located in the headwaters of the Thames. Both swamps, before their purchase in 1949, were susceptible to fires set to burn off the peat layer to facilitate farming. The Ellice Swamp, near Stratford, now covers 856 hectares and is now the largest forest in Perth County. It provides habitat for rare species such as the Red-headed Woodpecker and the Red-shouldered Hawk. The Dorchester Swamp, part of the Upper Thames east of London, has an important nesting heronry for the Great Blue Heron.[30]

Being located on more valuable "corn-belt-style" land than that of the Upper Thames Authority, the Lower Thames Conservation Authority endured even more political resistance to its formation. It would not be created until thirteen years later, in 1961, but resistance would have even been stronger had a single authority been proposed for the Thames. Once in place, it undertook reforestation in the Skunks' Misery area through the Mosa Forest in southwest Middlesex County and developed the forty-one hectare Ekrid Forest. The fact that one of its twenty-two conservation areas, Big Bend near Wardsville, is still partially under lease for farming demonstrates the continuing difficulties faced in encouraging reforestation.[31]

Despite formidable challenges, forest cover throughout southwestern Ontario is steadily, although slowly, increasing. The rate has been slowed by privatization of public nurseries in 1996, but increasingly restrictive tree-cutting bylaws have prevented the destruction of existing forests. However, contemporary conservation reports are documenting concerns regarding undersized Brook Trout in southwestern Ontario, identical to what Richardson found on the Ganaraska in 1943. There are other concerns: there needs to be at least a 15-metre forest buffer around open water to filter out contaminates from waters flowing into lakes — most streams in southwestern Ontario do not have this protection. The same problem contributes to *E. coli*-counts that are higher than provincial guidelines, making the water a health hazard. Yes, tiny Brook Trout in southwestern Ontario are exhibiting the same problems that were corrected in other parts of province by Zavitz's efforts. In other areas, there are no trout and streams are highly degraded. While Zavitz's legacy here has checked the worst flooding problems, streams are dried up much of the year and have little life in them. The small size of forests means they are quite vulnerable to winds and cannot provide habitat to forest interior birds like the Scarlet Tanager. Frog populations have dropped, and the list of issues continues. Little wonder the environmental commissioner has urged that Ontario plant a billion more trees.

Another achievement was the Catfish Creek Conservation Authority's use of the Agreement Forest Program in 1962 to acquire the 256-hectare White's Bush, now called Springwater Forest Conservation Area. According to the conservation authority's website, it has more endangered species than any natural area in Canada. Logged in an ecologically sustainable way prior to its purchase, the site has old-growth characteristics of tall White Pines and Tulip Trees. A testament to Zavitz's role in protecting Lake Huron watersheds in Southern Ontario is the fact that the threatened forests he investigated in his 1908 Wasteland Study are now the best-forested stretch in the Canadian "corn belt."

Established as provincial parks, these forested areas — together with forests on adjacent Indian reserves — provide the best forest cover for streams in the watersheds flowing into Lake Huron, whose headwaters suffer from extreme deforestation.[32]

For much of the time since the Second World War, Norfolk County's farmland grew very valuable crops, notably tobacco, although its profitability has declined in recent years. Despite this situation, Norfolk County has 30 percent forest cover, the best situation west of the Niagara Escarpment. To a large extent, this success is a credit to Zavitz, his circle of friends, and conservation authorities. Although 1,618 hectares had been reforested on public lands, primarily through the St. Williams and Turkey Point provincial initiatives as well as the Norfolk County forests, deforestation continued on private lands. A 1958 progress report for the Big Creek Conservation Authority documented that recently 320 hectares of forest in its watershed had been removed, thus increasing stream sedimentation. Such destruction on private lands included areas that had been reforested with the help of provincial foresters, often taking place "when the property changed hands."[33]

Those who sought to destroy Norfolk's forests, while powerful, had a determined adversary in Monroe Landon, a local farmer. In 1950 he left the job of running the family farm to his son Ken and became a paid, full-time tree-bylaw enforcement officer for Norfolk County. He carefully used the powers given to the county as part of the Trees Act that Zavitz persuaded Ontario Premier George Drew to pass in 1946. His sparse diary entries are largely records of his patrols. What stemmed the tide of illegal cutting was his securing a conviction in court of a farmer clear-cutting a forest in order to illegally increase his tobacco cultivation. Not daunted by a high-priced Toronto lawyer, Landon assured that the farmer was not only required to pay a fine but to reforest his property.[34]

Another victory for Norfolk conservationists in the 1950s was the extension of fire control to agricultural areas south of the Canadian Shield. Harold Zavitz, forester for the Lake Erie District, worked on the plan, which was accelerated in response to a 4,046-hectare fire on the Long Point peninsula in 1958. He later used it as a model for forest-fire response plans for Lambton and Middlesex Counties.[35] Using the legislative tools wrought by his cousin Edmund, Harold brought law and order into the forests of his Lake Erie District, which included most of the Carolinian region, including the Niagara Peninsula. Harold ended the pattern that Edmund had experienced of local timber barons like Cutler clear-cutting farm forests and littering what was left with fire-prone slash; in 1962 Harold introduced a field service to help farmers develop forest management plans for their properties. A periodic marketing bulletin was developed for farmers wishing to sell the timber on their properties.[36]

The Big Creek Conservation Authority became a new force for forest protection in Norfolk. By 1957, only six years after its creation, it had acquired "1,500 acres of forests," primarily swamp forests seen as "the natural storage areas." In 1959, it acquired the 282-hectare Backus Woods, which, much like White's Bush, had been logged sustainably and therefore retained old-growth characteristics. Big Creek's successor, the Long Point Conservation Authority, now protects 4,451 hectares of forests.[37]

Edmund Zavitz, in retirement, was in an excellent position to protect the forests of Norfolk County. His cousin Harold, the district forester, worked at St. Williams, just a few minutes drive away from his Forestville hobby farm. In Toronto he had a darkroom office in the Whitney Block, where he was able to organize and print many of his better pictures from the massive files of negatives he had assembled over the years. Here, he could also confer with his old friend the deputy minister Frank MacDougall over important matters of forest conservation. When working in Toronto, he stayed at his home in Brampton.

When Edmund was in retirement, Harold Zavitz was required to write a history of the Lake Erie District of the Department of Lands and Forests. Reading it provides some insights into what the two men would have talked about while the work was in process. In addition to discussions about extending fire control and extension services to private forest owners, the history provides details into matters such as the impact of the new provincial Wilderness Act of 1956 and the subsequent creation of Turkey Point Provincial Park in 1958. The Wilderness Act finally prevented provincial parks from being cut up into cottage lots, the disastrous practice that Edmund Zavitz had witnessed at Long Point in the early 1920s. Rondeau Provincial Park was also protected from further cottage encroachments at this time. In addition to Turkey Point, the act propelled the creation of a new chain of provincial parks along Lake Erie such as Rock Point, east of Dunnville.[38]

During his retirement, Edmund Zavitz spent many hours in the Whitney Block darkroom, and also used his spare time to write two books. The first, *Hardwood Trees of Ontario*, was published in 1959. The second, *Fifty Years of Reforestation in Ontario*, appeared in 1962. Another, more personal publication, "Recollections, 1875–1964," was published in 1965. His writings provided useful source material for the 1967 Centennial of the Department of Lands and Forest's publication, *Renewing Nature's Wealth*. This book remains the only comprehensive account of the historical development of provincial policies in Canada for the protection of the environment. Zavitz's account of reforestation in Ontario, while sparse, provides key details for the major figures. His is the only published book to mention the heroic figure behind reforestation in Eastern Ontario, Ferdinand Larose.

The most poignant of the three publications that Zavitz assembled while in retirement was *Hardwood Trees of Ontario*. The majestic giants of his photographs might be viewed as resembling the achievements of the author: just as these giants towered in height over the other trees of the forest, he towered above his contemporaries in terms of his impact on the protection of forests and his determined reforestation in the province of Ontario. The release of his book in 1959 came at an appropriate time — the towering sentinels of Rondeau were now protected from logging and cottage development, controlled by the zoning regulations of the Wilderness Act. Many of the key figures working with Zavitz throughout his lifetime of protecting forests and thus rescuing the province from ecological disaster, men like Judson Clark, Bernard Fernow, and J. H. White, like Zavitz himself, can be counted among the giants of the forest.[39]

Premier John Robarts planted the one billionth tree in Ontario in 1968, one month-and-a-half before Edmund Zavitz's death. Shown from left to right: Jim Drury, sixteen-year-old grandson of the late Honourable E.C. Drury; Premier John Robarts; Ross Zavitz, son of the late E.J. Zavitz; and the Honourable Rene Brunelle, minister of lands and forests. The figure of a billion trees illustrates that Zavitz's task of reforesting Ontario, although impressive, was only half completed. Zavitz's tripling of the forest cover in Southern Ontario required the use of a billion trees from the public nurseries he created. The fact that another billion trees is required to finish the job was detailed by the environment commissioner of Ontario, Gordon Miller, in his annual report to the Ontario legislature in 2010. He estimated that another billion trees would be needed to reach the bare minimum of 30 percent forest cover. The still heavily deforested regions, such as those in southwestern Ontario, currently only have 17 percent of the land in forest.

Zavitz's second wife Margaret died in 1964. He then lived with son Dean in Port Credit before going to a nursing home in 1967. Here, he would have derived great satisfaction in reading about his achievements in the pages of *Renewing Nature's Wealth*.

Just prior to Edmund Zavitz's death at the age of ninety-four, on December 30, 1968, in the Peel Memorial Hospital, the province of Ontario held a ceremony at Queen's Park to mark the planting of the billionth tree at the St. Williams Forestry Station. Premier John Robarts, a year earlier, had praised Zavitz as the only planner in "the development of sound, scientifically-based programs of conservation," who was "still living among us in ripe old age."[40]

Robarts had lived through and remembered the great flood of 1937 that devastated his hometown of London, Ontario. This disaster, which had destroyed 1,200 homes and killed five people, gave him some appreciation of how the work of the St. Williams Nursery, founded by Zavitz, played a major role in rescuing the province from ruin. Robarts's experience was reinforced after the Second World War by his service on London City Council, one of whose chief objectives was to establish a Thames River Conservation Authority. His early political experiences would have brought him into contact with such prominent figures as Watson Porter and Nelson Monteith, prompting the barrage of environmental initiatives his government introduced. These included the beginning of a Niagara Escarpment planning process, the Parkway Belt plan from Burlington to Toronto, and a vast increase in the number of provincial parks, including the creation of the enormous Polar Bear Provincial Park in the James Bay Lowlands. After Robarts's retirement, these policies were further developed by his conservationist-minded treasurer, John White, a fellow Londoner who also lived through the flooding disasters of the Depression.

The eulogy at Edmund Zavitz's funeral was delivered by the Reverend Milton Johnston of the Richmond Hill Baptist Church:

> That whosoever shall make two ears of corn or two blades of grass, to grow when only one grew before,

would deserve better of mankind and do more essential service to his country, than the whole race of politicians put together. Edmund Zavitz did not grow mere blades of grass, or sandy soils. He grew whole forests. In fact he was responsible for reforesting most of the once barren wastelands in Ontario.[41]

Dr. Edmund J. Zavitz was buried on January 2, 1969, a short distance from the Forestville farm in the Hillcrest Forestville Cemetery. Here, his wife Jessie and son John, who had predeceased him, are buried as well. Appropriately, the family grave marker is hugged by a Juniper bush.

As E.C. Drury pointed out, the life of Edmund Zavitz shows how humanity can redeem the ills inflicted on the earth, which bring forth disasters such as floods, landslides, forest fires, and marching deserts. While in other areas of the world such horrors are the staple of the television news, in Ontario they are mainly part of the history books. This is not only because of what Zavitz achieved in his lifetime but because of the ongoing work of various public institutions and advocacy groups that he and his friends created. Like his favourites, the gigantic White Pine and the Tulip Tree he so admired, Edmund Zavitz, in his long and productive lifetime, was a giant in his chosen field. Few individuals have contributed as much to Ontario.

— Appendix —

Chronology of the Life of Edmund Zavitz

1875, July 9: Edmund Zavitz born in Ridgeway, Ontario, then located in Bertie Township, Welland County, now part of Fort Erie, Regional Municipality of Niagara. His mother, Dorothy, and maternal grandfather Edmund Prout cultivate a love of trees and nature in young Edmund.

1887, September 25: Death of Zavitz's father, Joseph Zavitz.

1889: Completes Ridgeway Continuation School, where his interest in nature was encouraged by principal Alva Kilman. Goes to work at an uncle's farm, then to work for local business mogul, Eber Cutler.

1895: Mother and stepfather, I.L. Pound, convince him to return to high school. Enrols at St. Catharines Collegiate for grade 11. Finishes secondary school studies at Woodstock Collegiate.

1899: Attends McMaster University in Toronto, where he meets his future wife, Jessie Dryden, daughter of minister of agriculture, John Dryden.

1903: In last year of studies at McMaster, decides to become a forester and also becomes friends with Judson Clark, chief forester of Ontario. Clark teaches Zavitz photography on trips to see Niagara Escarpment Forests in Lincoln County and deserts in Norfolk County.

1903–04: Begins forestry studies at Yale University and completes his studies at University of Michigan. In summer of 1904, he establishes first provincial nursery at Ontario Agricultural College (OAC) in Guelph.

1905: Receives Master of Science in Forestry. In 1905 is hired as lecturer at OAC, marries Jessie Dryden. Establishes pilot tree-planting project on grandfather's farm, now owned by his relative Francis Squair. This is the beginning of a co-operative planting program with farmers on small parcels of their land.

1907–08: Teaches a course on dendrology at University of Toronto for one year on a part-time basis. Meets long-time friend and assistant, James Herbert White.

1908: Has OAC forestry station moved to St. Williams which he surrounds with a 445-hectare demonstration forest on the former sand dunes. He writes a wasteland report illustrated by his own photographs, which led to the Counties Reforestation Act of 1911. This legislation is not employed, however, by any county governments until 1921.

1912: Protests from Quebec and federal government over uncontrolled human-induced forest fires in northern Ontario result in Zavitz being appointed chief forester of Ontario. Initially, is only responsible for his previous OAC duties and protection from railway fires.

1917: His fire-control responsibilities expanded in 1917 as result of Great Matheson Fire of 1916. A permit system for burning by farmers in Northern Ontario is established, which is administered by fire rangers of his Forest Protection Branch.

1919–23: His close friend E.C. Drury becomes premier of Ontario for three-and-a-half years. With Drury's support, Agreement Forest Program develops, which provides assistance to municipalities undertaking reforestation projects. New nurseries opened at Midhurst and

Orono and the Ontario Tree Seed Plant at Angus. First Agreement Forest Project, the Hendrie Forest, is established near Midhurst. An Ontario Provincial Air Service is developed.

1923–34: Conservative governments under Howard Ferguson and George Henry gradually built on Drury's initiatives. All counties along the Oak Ridges Moraine take part in an Agreement Forest Program. Reforestation launched in Eastern Ontario under leadership of Ferdinand Larose. In 1926 the Department of Forests is created, with Zavitz as deputy minister. Its main achievement is to be able to convert forest reserves to provincial forests under control of professional foresters, and to be developed for multiple use, especially wilderness recreation. Eventually, most are protected as provincial parks. Threat to soils of northern forests by repeated burns ended by toughened regulations and effective use of his forest departments' Ontario Provincial Air Service.

1934: Shortly following death of Jessie and John Dryden Zavitz, wife and first son respectively, Edmund Zavitz loses control of Department of Forests as a result of election of Mitchell Hepburn as premier. Frederick Noad becomes deputy minister of forests and purges twenty key personnel of the Department of Forests created by Zavitz, including seventeen professional foresters and the director of the Ontario Provincial Air Service (Roy Maxwell was replaced by Hepburn's chauffeur). Zavitz becomes director of the Reforestation Branch, which effectively confines his duties to Southern Ontario.

1935–1946: Zavitz takes up issues of having tree-control bylaws to restrict cutting on private lands and the creation of conservation authorities, both of which are achieved through provincial legislation in 1946. He is aided in these efforts by a former OAC student Watson Porter. In 1941 the Department of Lands and Forests is reorganized along lines long advocated by Edmund Zavitz, through his long-time protege Frank MacDougall, now the deputy minister.

1949: Retires as director of the reforestation branch and for the next five years acts as consultant to the department.

1951: Receives Award of Merit from the Canadian Forestry Association.

1952: Receives Honorary Doctor of Law Degree from McMaster University.

1953: Retires from consultancy.

1957: Receives Honorary Doctor of Law Degree from the University of Toronto; Ontario Archaeological and Historic Sites Board plaque recognizing Zavitz as "the father of reforestation in Ontario" is erected at St. Williams, Norfolk County.

1959: Publishes *Hardwood Trees With Bark Characteristics*.

1961: Publishes *Fifty Years of Reforestation in Ontario*.

1964: Publishes "Recollections, 1985–1964."

1947–1968: During the last three decades of his life, Zavitz sees programs he advocated transform Ontario. Forest cover in Southern Ontario had tripled as a result of his reforestation efforts by the time of his death. In Northern Ontario, Crown lands firmly placed under professional foresters.

1968, December 30: Edmund Zavitz dies at age ninety-four and is buried in Hillcrest Forestville Cemetery, Norfolk County.

2011, June: Plaque erected at the University of Guelph to commemorate his years of involvement with Ontario Agricultural College, 1905–1912.

— *Chronology of the Life of Edmund Zavitz* —

2011, August 13/14: Edmund Zavitz was honoured at the annual Forest Fest, sponsored by the Port Rowan-South Walsingham Heritage Association and held at the St. Williams Nursery and Ecology Centre at St. Williams, Ontario. The St. Williams Conservation Reserve was officially designated as the Edmund Zavitz Forest.

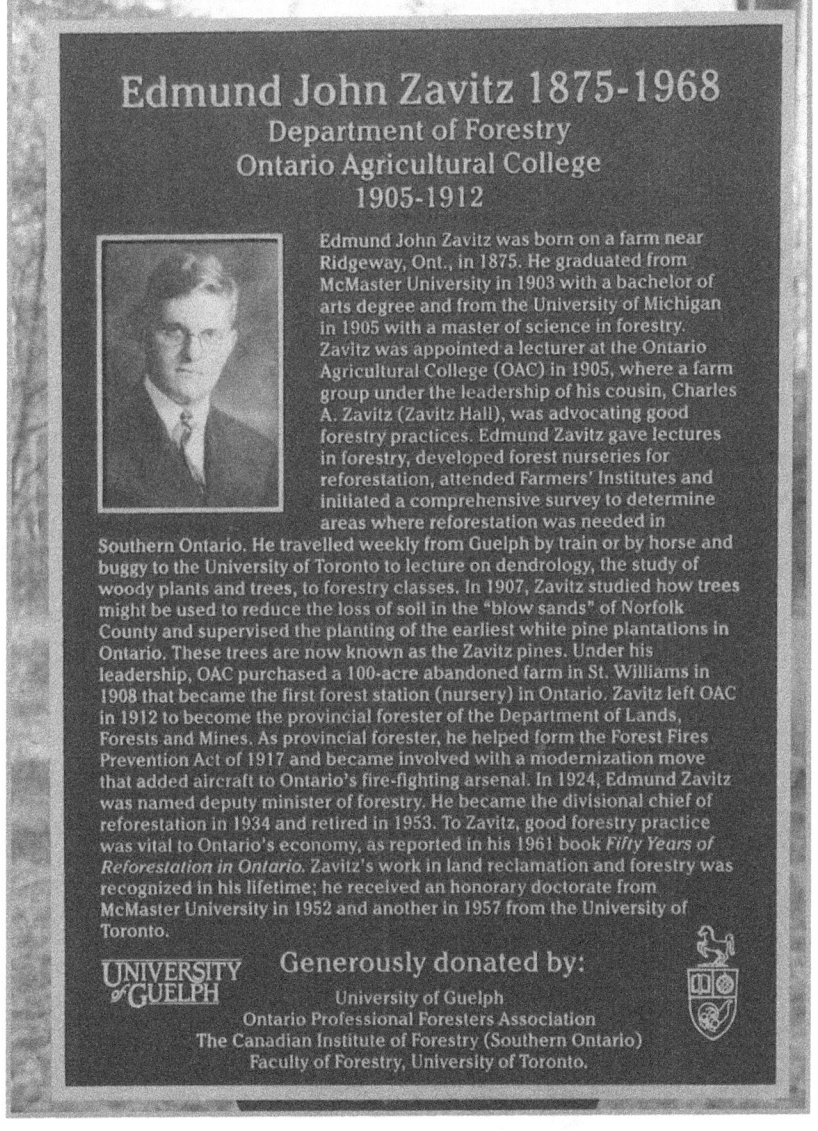

Plaque dedicated to Edmund John Zavitz, placed on the grounds of the University of Guelph on September 16, 2011.

Notes

Chapter One — Edmund Zavitz: The Man Who Did Plant Trees

1. Richard Lambert and Paul Pross, *Renewing Nature's Wealth* (Toronto: Department of Lands and Forests, 1967), 584.

2. John Robarts, "Introduction," *ibid.*, xvi.

3. "Extension Notes: Forest History of Eastern Ontario," Ontario Ministry of Natural Resources (1997), 7.

4. *Ibid.*, 21.

5. *Ibid.*, 7.

6. "Critical Review of Historical and Current Tree Planning Programs on Private Land in Ontario," Ontario Ministry of Natural Resources (2000), 7–9.

7. *Ibid.*, 16–17.; J.M. Buttle, "Hydrological Response to Reforestation in the Ganaraska River Basin, Southern Ontario," http://onlinelibrary.wiley.com/doi/10.1111/j.1541-0064.1994.tb01685.x/abstract (accessed May 12, 2011). Buttle, a geography professor at the University of Trent, examined the response of the Ganaraska River to basin reforestation and found it had an even better result in reducing flash runoff and increasing water storage

in headwater aquifers than had been predicted by supporters of reforestation.

8. Char Miller, *Gifford Pinchot and the Making of Modern Environmentalism*, (Washington: Island Press, 2004), *passim*.

9. Laurence C. Walker, *Forests: A Naturalist's Guide to Trees and Forest Ecology*, (Toronto: John Wiley, 1991), 99. Gifford Pinchot also reforested the lands around his family estate Grey Towers. His early efforts at protecting White Pine are in vivid contrast to the hotly contested grounds of Oka. According to Aubrey White (*History of Forest Regulations in Ontario*), many apologists for late nineteenth-century forest exploitation claim that White Pine were inevitably doomed.

10. Peter MacKay, *Heritage Lost: The Crisis in Canada's Forests* (Toronto: Macmillan of Canada, 1985), 98.

11. J.R.M. Williams, "Southern Ontario Newsletter," *Canadian Institute of Forestry* (Spring 2006). A professional forester, Mack Williams' entire career was with the Ontario Department of Lands and Forest/MNR. He is best known for his work in reforestation, for which he received the White Pine Award from the Ontario Forestry Association and the Presidential Award from the Canadian Institute of Forestry in 2007. J.M.R. Williams died in January 2011.

12. Peter MacKay, *Heritage Lost*, 98.

13. Mack Williams, "Southern Ontario Newsletter."

Chapter Two — Early Influences

1. Fort Erie Museum, "Many Voices 11," Fort Erie Historical Board (2004), 365.

2. *Ibid.*, 363.

— *Notes* —

3. After five years of debating the forestry issues on Welland County Council, at one point underscored by his dramatic resignation from its conservation committee, Swart successfully pushed through Welland County's first tree-cutting bylaw. He persevered after being encouraged by the news that such statutes could actually work. Monroe Landon, the enforcement officer in Norfolk County with the authority to charge an individual engaged in illegal clear cutting, had successfully concluded a prosecution for such an offence. Following this victory, Swart concentrated his efforts on establishing the Niagara Peninsula Conservation Authority, which came into being in 1959. Information gathered in a personal interview with Mel Swart.

4. Edmund Zavitz, "Recollections, 1875–1964," Department of Lands and Forests (1964), 2; John Macfie, "A Picture Needs a Thousand Words," Archives of Ontario, hereafter AO, *www.archives.gov.on.ca*.

5. John Squair, "History of Darlington and Clarke Townships," (Toronto: University of Toronto Press, 1927), 3.

6. "Our 160 Year Old Forester, Edmund Zavitz," *Sylva*, Vol.1, No.1, Department of Lands and Forests, undated copy, located in the Edmund Zavitz file, Fort Erie Museum. The phrase "160 year old forester" is taken from an error in the birthday on Zavitz's University of Michigan degree.

7. John Squair was a professor of French at the University of Toronto. He was decorated by the French government through its highest award, the Legion of Honour, to commend his advocacy of elementary school teaching of French in Ontario schools in order to nurture better relations between Ontario and Quebec. Donald Jones, "U. of T. Professor Received France's Highest Honour," *Toronto Star*, March 5, 1983.

8. Squair, "History of Darlington and Clarke Townships," 3.

9. John Squair Papers, University of Toronto Archives, B77-0027.1001.

10. Squair, "History of Darlington and Clarke Townships," 10.

11. Ibid., 9.

12. Ibid., 7–10.

13. Bertie Township School, S.S. No. 11, now a retail complex, is still standing in Ridgeway, a few blocks from the Zavitz home.

14. Fort Erie Museum, "Many Voices 11," 143, 152–153, 206, 299, 333.

15. Zavitz, "Recollections," 3–4.

16. Ibid.

17. Fort Erie Museum, "Many Voices 11," 143–53.

18. Alexander Ross and Terry Crowley, *The College on the Hill*, (Toronto: Dundurn Press, 1995), 30–50; Charles M. Johnson, *E.C. Drury: Agrarian Idealist*, (Toronto: University of Toronto Press, 1997), 18–22.

19. Ross and Crowley, *The College on the Hill*, 30–50.

20. Zavitz, "Recollections," 3; Department of Lands and Forests, "Our 160 Year Old Forester," 3.

21. McMaster University opened in Toronto in 1887, established by funds bequeathed by the Honourable William McMaster. The university was relocated to Hamilton in 1930.

22. Ian Stuart, "John Dryden," *Canadian Dictionary of Biography*, Volume XIII, 1901–1911, (Toronto: University of Toronto Press, 1994), 287–89.

23. Ibid., 3–4

24. "Our 160 Year Old Forester," 4.

— Notes —

25. "Forestry expert, native of Ridgeway to be Honoured," unidentified newspaper clipping, Fort Erie Museum, Edmund Zavitz file.

26. "Our 160 Year Old Forester," 4.

27. Press clipping from the *Barrie Free Herald Press*, January 7, 1948, AO, F7-MU, Ernest Drury Papers.

28. *Ibid.*

29. *Ibid.*

30. "Our 160 Year Old Forester," nk.

31. "Forestry expert, native of Ridgeway to be Honoured," Fort Erie Museum.

32. Mayoral Proclamation and newspaper clipping, *Niagara Falls Review*, May 25, 1970, Fort Erie Museum, Edmund Zavitz File.

33. Letter, dated July 13, 1975, from Edith Jackson to one of Edmund Zavitz's surviving sons, Fort Erie Museum, Edmund Zavitz File.

34. Communication to author from Kathleen Mackenzie, January 22, 2011.

Chapter Three — Behind the Scenes

1. Sally M. Weaver, "The Iroquois and the Consolidation of the Grand River Reserve in the Mid-Nineteenth Century," in Edward Rogers and Donald Smith, eds., *Aboriginal Ontario* (Toronto: Dundurn Press, 1994), 197.

2. Stephen J. Pyne, *Awful Splendour: A Fire History of Canada* (Vancouver: University of British Columbia Press, 2007), 102; Arthur Lower, *Great Britain's Woodyard* (Montreal: McGill-Queen's University Press, 1973) *passim*.

3. Harold C. Zavitz, "A History of the Lake Erie Forest District," the Ontario Department of Lands and Forests (1963), 8.

4. Kenneth Kelly, "Damaged and Efficient Landscapes in Rural Ontario, 1888–1907," *Ontario History*, Vol. 67, No. 1 (March 1974), 8.

5. C.F. Coons, "Reforestation of Private Lands in Ontario," Forestry Research Group, *Armson Private Land Forestry Review* (1981), 5.

6. *Ibid.*, 5–20.

7. William Saunders (1836–1914) developed new fruit varieties for cold hardiness to expand the range of fruit growing in Canada. He founded both the Entomological and Canadian Pharmaceutical Societies. As well, Saunders established the Canadian government's network of Experimental Farms, and personally selected their locations. The Central Experimental Farm in Ottawa was designated as a National Historic Site in 1998.

8. "Reforestation of Private Lands in Ontario," 8–10.

9. Richard S. Lambert and Paul Pross, *Renewing Nature's Wealth*, 112, 162–163, 178, 224, 525–26.

10. Andrew Denny Rodgers, *Bernard Eduard Fernow: A Study of North American Forestry* (Princeton: Princeton University Press, 1950), *passim*.

11. Kelly, "Damaged and Efficient Landscapes in Rural Ontario," 5–8.

12. Ken Armson, *Ontario Forests*, (Toronto: Fitzhenry and Whiteside, 2006), 106.

13. *Ibid.*, 130–33.

14. The Commission of Conservation was a federal-provincial body initiated in 1909 by the Canadian prime minister Sir Wilfrid Laurier. It was abolished during the short-lived government of Prime Minister Arthur Meighen in 1921. Among its seven committees was a Forestry

Committee. Zavitz's close friend, J.H. White, served as its secretary. During its lifetime, it produced a number of important books, reports, and scientific papers. It also published a provocative newsletter, "Conservation." Its studies were attractively bound and illustrated, proving to be persuasive to readers. The commission carefully documented important threats to forests in Canada and demonstrated how forest soils in the Canadian Shield were being destroyed by repeated forest fires — not enough seed trees were being left to secure regeneration of valuable pine forests. Its work also gave publicity to Zavitz's efforts to reforest sand wastelands and to curb forest fires.

15. Commission of Conservation, Committee on Forests, *Trent Watershed Survey*, (Toronto: Bryant Press, 1903), *passim*.

16. Coons, "Reforestation of Private Lands in Ontario," 5–10.

17. Kelly, "Damaged and Efficient Landscapes in Rural Ontario," 5–8.

18. Ian Stuart, "John Dryden," *Canadian Dictionary of Biography*, Volume X111, 287–90.

19. Charles M. Johnson, "*E.C. Drury: Agrarian Idealist*," (Toronto: University of Toronto Press, 1991), 18–22.

20. *Ibid*.

21. W. Stafford Johnston and Hugh M. M. Johnston, *History of Perth County to 1967* (Stratford, ON: County of Perth, 1967), 343–44.

22. Roland D. Craig, "Co-operative Experiments in Forestry," in "The Report of the Department of Agriculture, 1903," Ontario Sessional Papers (1903), 37.

23. T.H. Mason, quote in "Report of the Experimental Union, 1903," Ontario Sessional Papers (1903), 35–50.

24. Craig, "Co-operative Experiments in Forestry," 39–46. Includes the next nine consecutive quotes taken from this document.

25. Charles M. Johnson, *E.C. Drury: Agrarian Idealist*, 18–22.

26. Zavitz, "Recollections," 4–5.

27. Squair, "History of Darlington and Clarke Township," 3.

28. Zavitz, "Recollections," 4.

29. David Dempsey, *Ruin and Recovery: Michigan's Rise and a Conservation Leader* (Ann Arbor: University of Michigan Press, 2002), 8.

30. Zavitz, "Recollections," 5.

31. *Ibid.*

Chapter Four — Exiled to Agriculture

1. Lambert and Pross, *Renewing Nature's Wealth*, 185. Frank Cochrane (1852–1919), was born in Quebec. A businessman, he owned a chain of hardware stores in Northern Ontario (Cochrane Hardware Limited) and had extensive mining and hydro development interests in the north. Although he lived in Southern Ontario, his political riding was Nipissing East. Cochrane did not like regulations he perceived as slowing northern growth, and he distrusted foresters for advocating them. He was quite outraged, for instance, when B.E. Fernow wrote a report for the Commission of Conservation that was critical of farming efforts in the Clay Belt.

2. Zavitz, "Recollections," 6.

3. *Ibid.*; E.C. Drury, *Farmer-Premier* (Toronto: University of Toronto Press, 1965), 54–55; Charles M. Johnston, *E.C. Drury: Agrarian Idealist*, 22–23.

4. Edmund J. Zavitz, "Report on Reforestation of Waste Lands of Southern Ontario," the Ontario Department of Agriculture (1908), 5.

— Notes —

5. Zavitz, "Recollections," 6; Springwater 50th Anniversary (2008), "Parks Blog Archives, Springwater 50th Anniversary," http://parkreports.com/.

6. Andrew Denny Rodgers, *Bernard Edward Fernow: A Study of North American Forestry* (Princeton: Princeton University Press, 1950), 250–300; Zavitz, "Recollections," 8; Gerald Killian, *Protected Places: A History of Ontario's Provincial Park System* (Toronto: Dundurn Press, 1993), 42; E.J. Zavitz, "Hardwood Trees of Ontario With Bark Characteristics," (Toronto: Ministry of Natural Resources, 1973), 22, 24, 43, 46, 56.

7. Zavitz, *Hardwood Trees of Ontario*, 20; Michael Henry and Peter Quirby, *Ontario's Old-Growth Forests* (Toronto: Fitzhenry & Whiteside, 2010), 66, 70, 104.

8. Zavitz, "Recollections," 8; Michael Barnes, *Killer in the Bush: The Great Fires of Northwestern Ontario*, (Erin Mills: Boston Mills Press, 1987), 30.

9. Edward Sisan, *Forestry Education at Toronto*, (Toronto: University of Toronto Press, 1965), 40–60.

10. Ken Armson, *Ontario Forests* (Toronto: Fitzhenry & Whiteside, 2000), 187.

11. Harry Barrett, *They Had a Dream: A History of the St. Williams Forest Station* (Port Rowan, ON: South Walsingham Heritage Association, 2008), 64.

12. Dean C.D. Howe to William Finlayson, January 4, 1928, AO, RG1, A-1-10, Forestry Branch.

13. E.J. Zavitz, "Report on Forestry for Ontario For 1906," part of "Report of the Experimental Union," 1906, Ontario Sessional Papers (1906), 39.

14. *Ibid.*, 40–43.

15. *Ibid.*, 42.

16. E. J. Zavitz, "Reforestation in Ontario," pamphlet reprinted by the Department of Lands and Forests. The original article appeared in *Canadian Geographic* (April 1947), 7; Alan Watson, "Zavitz Pines: One Hundred Years of Growth," *The Green Web* (Spring 2000) 1, 4.

17. Edmund Zavitz, "Farm Forestry," in *Ontario Agricultural College Bulletin 155*, (1987), 7.

18. *Ibid.*, 15.

19. Zavitz, "Report on Reforestation of Waste Lands of Southern Ontario," (1908), 10–14.

20. *Ibid.*, 10, 11; personal communication by Dolf Wynia. Many of Ontario's savannahs were created by Native land management techniques for such increasing populations of game. This means that to perpetuate such ecosystems, and the rare species that depend upon them (i.e. Lupines and the Karner Blue Butterflies they support), it is necessary to have controlled burns to modify what in the absence of such management, would become a mature, canopy forest.

21. Barrett, *They Had A Dream*, 30–37.

22. Zavitz, "Report on the Reforestation of Wastelands of Ontario," (1908), 6–10.

23. *Ibid.*; Ken Armson, W. Ross Grinnel, and Fred Robinson, "History of Reforestation in Ontario," in Robert G. Wagner and Stephen Colombo, eds., *Regenerating the Canadian Forest* (Toronto: Fitzhenry & Whiteside, 2001), 6.

24. Barrett, *They Had A Dream*, 30;

25. *Ibid.*, 30.

26. *Ibid.*, 29–33.

27. *Ibid.*, 36–37.

28. *Ibid.*, 70–75.

29. *Ibid.*, 70.

30. Zavitz, "Report on the Reforestation of Wastelands of Ontario," (1908?), 19–20.

31. Clyde Levitt, *Forest Protection in Canada, 1912* (Toronto: Commission of Conservation, 1912), 124; "Department of Planning and Development: Moira Conservation Report, 1965," Department of Planning and Development (1965), 71–72.

32. Edmund Zavitz, "Report on Co-operative Forestry," in "Report of the Experimental Union, 1907," Ontario Sessional Papers (1907), 5.

33. Ken Drushka, *H.R.: A Biography of H.R. MacMillan* (Madera Park, BC: Harbour Publishing, 1995), 36.

34. Zavitz, "Recollections," 7.

35. Barnes, *Killer in the Bush*, 30–31.

36. Donald MacKay, *Heritage Lost* (Toronto: MacMillian of Canada, 1985), 55.

37. Patrick Blanchet, *Forest Fires: The Study of War* (Montreal: Cantos, 2005), 58–90.

38. Stephen Pyne, *Awful Splendour: A Fire History of Canada* (Vancouver: University of British Columbia Press, 2007), 188.

39. Zavitz, "Recollections," 7.

40. Undated Memorandum to Minister of Lands and Forests, William Hearst, AO, William Hearst Papers, MU 309.

41 When, following the 1916 Matheson Fire, Zavitz was given the authority to investigate provincially chartered railways, he found,

"In all, 711 locomotives were inspected during the summer, of which, 229, or 30 per cent, showed defective screens, ash pans, or other appliances." Edmund Zavitz, "Report of the Forestry Branch," Ontario Sessional Papers (1916–17), 150.

Chapter Five — The Struggle Against Indifference

1. Address by E.C. Drury on "Reforestation," AO, Ernest Charles Drury Papers, F7-MU 955.

2. Memorandum, February 17, 1916, from Deputy Minister Alan Grigg to Howard Ferguson, AO, Forestry Branch Correspondence Files, 104-MTL.

3. Edmund Zavitz, "Report on the Forestry Branch, 1908," in "Report of the Department of Lands and Forests, 1908," Ontario Sessional Papers (1908), 155.

4. Edmund Zavitz, "Memorandum to William Hearst," November 6, 1913, AO, William Hearst Papers, F6-MU 1311.

5. Address to Experimental Union by George Creelman, in "Report of the Experimental Union, 1912," Ontario Sessional Papers (1913), 55–56.

6. Charles M. Johnston, *E.C. Drury: Agrarian Idealist* (Toronto: University of Toronto Press, 1996), 75.

7. Clyde Leavitt, *Forest Protection in Canada, 1912* (Toronto: Bryant Press, 1913), 133.

8. Edmund Zavitz, "Memorandum to William Hearst," AO, Sir William Hearst Papers, F6-MI, 131.

9. D. Strickland, *Trees of Algonquin Park* (Whitney: Ontario Ministry of Natural Resources and the Friends of Algonquin Park, 1989) *passim.*

— *Notes* —

10. Zavitz, "Report of the Forestry Branch, Department of Lands and Forests, 1913–1914," Ontario Sessional Papers (1915), 101.

11. Zavitz, "Report of the Forestry Branch, Department of Lands and Forests, 1913–1914," 101–03.

12. "Report of the Superintendent of Algonquin Park, 1914-15," in "Department of Lands and Forests, Annual Report, 1914–15," Ontario Sessional Papers (1916), 64–66.

13. Edmund Zavitz, "Report of the Forestry Branch, 1904–15," 66.

14. D'Arcy Jenish, "The Trail Blazers," www.ontarionature.org/discover/PDFs/trailblazers.pdf

15. Richard Lambert and Paul Pross, *Renewing Nature's Wealth*, 293–94; Joan Murray, *Tom Thomson's Trees* (Toronto: MacArthur & Co., 1999), 1–16.

16. For more information on Tom Thomson, see Roy MacGregor, *Northern Light: The Enduring Mystery of Tom Thomson and the Woman Who Loved Him* (Toronto: Random House Canada, 2010).

17. Jim Poling, *Tom Thomson*, (Canmore, AB: Altitude Publishing. 2003), 100–03. Unlike other commentators on Thomson's death, who lack his understanding of environmental history, Poling is aware of the murders in this era of wardens employed by the American Audubon Society. Should Poling's suppositions be wrong, and the explanation of his death being caused by a money dispute is true, or even another rationale, the work illustrates the significance of Zavitz's concern that fire rangers be better paid.

18. Edmund Zavitz, "Report of the Forestry Branch, 1912–13," 164.

19. Strickland, *Trees of Algonquin Park*, passim: R.C. Hosie, *Forest Regeneration in Ontario* (Toronto: University of Toronto Press, 1953), 293-94.

20. Edmund Zavitz, "Report of the Forestry Branch, 1913–1914," 85–86; Edmund Zavitz, "Report of the Forestry Branch, 1912–1913," Department of Lands and Forests Annual Report, 1912–1913," Ontario Sessional Papers (1914), 96–103.

21. Edmund Zavitz, "Report of the Forestry Branch, 1914–1915," 85–89.

22. Michael Barnes, *Killer in the Bush: The Great Fires of Northeastern Ontario* (Erin Mills, ON: Boston Mills Press, 1987), 56.

23. Zavitz, "Report of the Forestry Branch, 1912–1913," *passim*; Lambert and Pross, *Renewing Nature's Wealth*, 205, 220, 235, 241.

24. Zavitz, "Report of the Forestry Branch, 1914–1915," 85–90.

25. Barnes, *Killer in the Bush*, 55–63.

26. Poling, *Tom Thomson*, 100–03.

27. Stephen Pyne, *Awful Splendour: A Fire History of Canada*, (Vancouver; University of British Columbia Press, 2006), 424.

28. Barnes, *Killer in the Bush*, 62–63.

29. Edmund Zavitz, "Recollections," 12.

30 *Ibid.*

31. Clyde Leavitt, "Report of the Committee on Forests, Eighth General Meeting," Commission of Conservation (January 16–17, 1917), 137–38.

32. *Ibid.*

33. *Ibid.*

34. Edmund Zavitz, "Report of the Forestry Branch, 1919–1920," "Annual Report of the Department of Lands and Forests, 1917–18," Ontario Sessional Papers (1918), 108.

35. *Ibid.*, 109.

36. *Ibid.*

37. *Ibid.*, 117.

38. *Ibid.*, 108.

39. *Ibid.*, 119–20.

40. Lambert and Pross, *Renewing Nature's Wealth*, 232.

41. *Ibid.*, 162.

Chapter Six — Drury and Zavitz: A Partnership

1. Ken Armson, W. Ross Grinnel, and Fred Robinson, "History of Reforestation in Ontario," in Robert W. Wagner and Stephen Colombo eds., *Regenerating the Canadian Forest* (Markham, ON: Fitzhenry & Whiteside, 2001), 3.

2. Speech by E.C. Drury on "Reforestation," August 11, 1937, AO, Ernest Charles Drury Papers, F79MU, 955, Box 6.

3. *Ibid.*

4. Harold C. Zavitz, "History of the Lake Erie District," Department of Lands and Forests (1963), 15–20.

5. Zavitz, "Report of the Forestry Branch, 1920," 215–220.

6. Zavitz, "Report of the Forestry Branch, 1919," in *Report of the Department of Lands and Forests, 1920*, Ontario Sessional Papers (1920), 108–09.

7. Bruce M. Pearce, "Norfolk's Reforestation Story: The St. Williams Forestry Station," in *Historic Highlights of Norfolk County* (Port Dover, ON: self-published, 1973), 215–20.

8. E.C. Drury, *Farmer-Premier* (Toronto: McClelland & Stewart, 1966), 99.

9. Zavitz, "Report of the Forestry Branch, 1920," 358.

10. Zavitz, "Report of the Forestry Branch, 1922," 197.

11. Personal interview with Brain Swaile, former manager of the Ontario Seed Tree Plant; Oak Ridges Moraine Foundation, www.moraineforlife.org.

12. Zavitz, "Report of the Forestry Branch, 1920," 358.

13. Zavitz, "Report of the Forestry Branch, 1920," 215–20.

14. Lambert and Pross, *Renewing Nature's Wealth*, 208.

15. Zavitz, "Report of the Forestry Branch, 1920," 200.

16. Squair, *History of Clarke and Darlington Township*, 12.

17. Zavitz, "Report of the Forestry Branch, 1919," 118.

18. Zavitz, "Report of the Forestry Branch, 1922," 197.

19. C.F. Coons, "History of the Reforestation of Private Lands in Ontario," a booklet published in Toronto by Armson and Associates (1909), 20–30; Newspaper Clippings, AO, Forestry Branch, Series RG1-560, Clipping *Ottawa Citizen*, "Ontario Forestry Field Days."

20. Zavitz, "Report of the Forestry Branch, 1922," 193–94.

21. Drury, "Speech on Reforestation," 1937.

22. Drury, *Farmer-Premier*, 13.

23. Zavitz, "Recollections," 8.

24. John C. Carter, "Ontario Conservation Authorities: Their Heritage Resources and Museums," *Ontario History*, Vol. 88, No. 2 (Spring

2002), 10–15.

25. Mark Kuhlberg, "Ontario's Nascent Environmentalists: Seeing the Forests for the Trees in Southern Ontario, 1919–1929," *Ontario History*, Vol. 94, No. 1 (June 1996), 124–29.

26. Zavitz, "Report of the Forestry Branch, 1921," 239–40.

27. Zavitz, "Report of the Forestry Branch, 1933," 190: "Simcoe County Forests from Past to Present," county.simcoe.on.ca.

28. "Simcoe County Forests From Past to Present"; Ed Borczon, *Evergreen Challenge: The Agreement Forest Story* (Toronto: Ministry of Natural Resources, 1982), *passim*.

29. Edmund Zavitz, "Report of the Forestry Branch, 1922," 252; Brenlee Robinson, "From Blowsands to Mixed Forest: The History of the York Region Forest," unpublished M.A. thesis, University of Toronto School of Forestry (2005), 13–14.

30. "Sandbanks Management Draft Master Plan," Ontario Parks website, www.ontarioparks.com.

31. Stephen Pyne, *Awful Splendour: A Fire History of Canada* (Vancouver: University of British Columbia Press, 2007), 254.

32. Zavitz, "Recollections" 17.

33. Richard St. Barbe Baker, "Richard St. Barbe Baker Remembers Men of the Trees — Diamond Jubilee," in *Trees Sixty Years Towards the Future* (Crowley Down, UK: The International Society for Planting and Protection of Trees, 1983), 16–23.

34. Zavitz, "Report of the Forestry Branch, 1923," 169.

35. Johnston, *E.C. Drury, Agrarian Idealist*, 203, 206.

36. Robert Roswell Gamey, Newspaper Clippings File, University of Toronto Archives, Graduate Records, Bernard Fernow.

37. Johnston, *E.C. Drury, Agrarian Idealist*, 175.

38. Paul Pross, "The Development of Professions in the Public Service: The Foresters of Ontario," *Canadian Public Administration*, Volume 10 (1967), 389–90. Many of the gains that Pross records in the struggles of foresters were reversed in the mid-1980s; the role of making administrative decisions having been assigned to non-scientist managers.

39. Judson Clark, "Report," August 12, 1922, in "Annual Report of the Department of Lands and Forests, 1922," Ontario Sessional Papers (1922–23), 275–79.

40. *Ibid.*, 277–79.

41. Zavitz, "Report of the Forestry Branch, 1923," 173.

42. Zavitz, "Recollections," 14.

43. *Ibid.*

44. Michael Barnes, *Killer in the Bush* (Eden Mills: Boston Mills Press, 1986), 98.

45. Memorandum from Edmund Zavitz to Beniah Bowman, AO, Forestry Branch, Correspondence Files, Revision Forest Fire Prevention Act, RG-1-256.

46. E. C. Drury, *Farmer-Premier*, 140–200.

47. Barnes, *Killer in the Bush*, 98.

48. Zavitz, "Recollections," 14.

49. Fred McClement, *The Flaming Forest* (Toronto: McClelland & Stewart, 1969), 108.

50. Zavitz, "Report of the Forestry Branch, 1922," 221.

51. Zavitz, "Recollections," 14.

— Notes —

52. Zavitz, "Report of the Forestry Branch, 1937," 130.

53. "History of Cochrane District," Ontario Department of Lands and Forests (1964), 10–15.

54. E.C. Drury, "Speech on Reforestation," delivered at the founding meeting of the Ontario Conservation and Reforestation Association in Barrie, Ontario, in 1937.

Chapter Seven — A Decade of Environmental Reform

1. Paul Pross, "The Development of Professions in the Public Service: The Foresters of Ontario," *Journal of Canadian Public Administration*, Vol. 10 (1967) 30–35.

2. Gerald Killan, *Protected Places* (Toronto: Dundurn Press and Ontario Ministry of Natural Resources, 1993), 27.

3. Zavitz, "Recollections," 11.

4. *Ibid.*, 17.

5. Zavitz, "Report of the Department of Forests, 1928," 22–23.

6. "Turkey Point Provincial Park," of Tall Grass Ontario: "Turkey Point," ontarioparks.com.

7. Harry Barrett, *They Had a Dream*, 50–60.

8. Zavitz, "Report of the Department of Forests, 1930," 141–42.

9. Zavitz, "Report of the Department of Forests, 1928," 142–43.

10. Barrett, *They Had a Dream*, 69.

11. Zavitz, "Report of the Department of Forests, 1933," 107.

12 Barrett, *They Had a Dream*, 69.

13. Zavitz, "Report of the Department of Forests, 1934," 108.

14. Winifred Cains Wake *et al*, *Nature Guide to Ontario* (Toronto: University of Toronto Press, 1997), 99; Brenda Robinson, *From Blow Sand to Mixed Forest: The History of the York Regional Forest* (Toronto: University of Toronto School of Forestry, 2005), 13–14; Smart Wood, "Forest Management Plan For York Region Forest," Summary of Recommendations, www.smartwood.org.

15. "Our Forest: Our Future: Dufferin County Forest Management Plan," Background Report, 1995–2005, Dufferin County, *passim*.

16. Zavitz, "Report of the Department of Forests, 1928," 33, 36.

17. Zavitz, "Report of the Department of Forests, 1934," 105.

18. "Why the County Has a Forest," Northumberland County, Northumberlandcounty.ca.

19. Ottawa Field Naturalist website tribute to Ferdinand Larose, ofric.ca.

20. "Larose Forest: Ontario Forest of the Year," Ministry of Natural Resources (1989), *passim*.

21. Ferdinand Larose, "The South Nation and Its Environs," in *Conservation in Eastern Ontario* (Toronto: Department of Lands and Forests, 1947), 107–08.

22. *Ibid.*, 100.

23. *Ibid.*, 109.

24. *Ibid.*, 107–111.

25. Ottawa Field Naturalists, BioBlitz in Larose Forest 2007, Ottawa Field Naturalists, ofric.ca.

26. Lester H. Crosbie, "Report of the Tweed District," September 28, 1928, University of Toronto Archives, J.H. White Papers, Box 2.

— *Notes* —

27. J.H. White, "Forest Research in Ontario," University of Toronto Archives, J.H. White Papers, Box 2.

28. Zavitz, "Report of the Department of Lands and Forests, 1934," 107.

29. Adrian Hayes, *Parry Sound: Gateway to Northern Ontario* (Toronto: Natural Heritage Books, 2005), 158–59.

30. *Ibid.*, 189.

31. White, "Forestry Research in Ontario," *passim*.

32. Zavitz, "History of Reforestation in Ontario," 20.

33. *Ibid.*, 20–23.

34. White, "Forestry Research in Ontario."

35. *Ibid.*

36. Lester H. Crosbie to J.E. Sharpe, February 29, 1930, AO, Forestry Branch Correspondence File 104-MTL, Timber Leases.

37. Letter dated December 12, 1930, from Peter McEwen to E.J. Zavitz, AO, Forestry Branch Correspondence File 104, MTL, Timber Leases.

38. Bruce West, *Firebirds* (Toronto: Ministry of Natural Resources, 1977), 173–77.

39. *Ibid.*, 177.

40. Gerald Killian, *Protected Places*, 59–74.

41. Robert Pike, "Hell and High Water," *American Heritage Magazine*, Vol. 18, No. 2 (February 1967), 3; Mark Kuhlberg, *One Hundred Rings and Counting: Forestry Education in Toronto and Canada, 1907–2007* (Toronto: University of Toronto Press, 2009), 100.

42. Zavitz, "Recollections," 21.

Chapter Eight — From Disaster to Triumph

1. Richard Lambert with Paul Pross, *Renewing Nature's Wealth*, 229–230, 326, 328.

2. *Ibid.*, 208–09.

3. Edmund Zavitz, "Testimony to the Legislative Committee on the Department of Lands and Forests," *Journal of the Legislative Assembly of Ontario* (1941), Appendix One, 331.

4. Ken Armson, W. Ross Grinnell, and Fred C. Robinson, "History of Reforestation in Ontario," in Robert G Wagner and Stephen J. Colombo, *Regenerating the Canadian Forest* (Toronto: Fitzhenry & Whiteside, 2001), 9.

5. Mark Kuhlberg, *One Hundred Rings and Counting* (Toronto: University of Toronto Press, 2009), 126.

6. Lambert and Pross, *Renewing Nature's Wealth*, 343, 346–347, 353, 529, 530.

7. Zavitz, "Recollections," 10.

8. A.H. Richardson, *Conservation for the People* (Toronto: University of Toronto Press, 1974), 3–4.

9. AO, Forestry Branch Newspaper Clippings Book, RG-18, 125, Editorial Watson Porter, *The Farmers' Advocate*, May 12, 1938.

10. Websites of Thomas Talbot Land Trust and International Biological Inventory, Skunk's Misery Complex, Lower Thames Conservation Authority, Mosa Forest, www.lowerthames-conservation.on.ca.

11. AO, Ernest Charles Drury Papers, Newspaper Clippings, MU 955.

12. Richardson, *Conservation for the People*, 12.

13. AO, Forestry Branch Newspaper Clipping Book, Watson Porter, *The Farmers Advocate*, nd., RG-18-125.

14. *Ibid.*, Newspaper clipping from *Ottawa Citizen*, September 28, 1938.

15. Management Plan for the Lanark County Forest, Lanark County, www.county.lanark.on; Management Plan For the Limerick Forest, United Counties of Leeds and Greenville, www.uclg.ca.

16. "Forests" Grey County, www.grey.ca.

17. Copy of letter to Edmund Zavitz, June 5, 1939, in the J.H. White Papers, University of Toronto Archives, B83-00/008.

18. Bird census of the David Dunlap Observatory, Website Richmond Hill Field Naturalists, www.rhnaturalist.ca. Like many of Zavitz's forestry achievements, the forest around the David Dunlop Observatory is threatened by residential development. At the time of this writing, the Richmond Hill Field Naturalists are trying to protect it.

19. John Bacher, "The History of Coronation Park," *Urban History Review*, Vol. 19, No. 3 (February 1991), 210–17.

20. *Ibid.*, 204–05.

21. Zavitz, "Recollections," 23.

22. *Ibid.*

23. *Ibid.*

24. Zavitz, "Testimony to the Legislative Committee on the Department of Lands and Forests," Journal of Legislative Assembly of Ontario (1941), Appendix One, 306–311, 327. The journal also contains testimony of John Irwin and Minister Peter Heenan.

25 Ibid., 306.

26. John C. W. Irwin, "Testimony to the Legislative Committee on the Department of Lands and Forests" (1941), 732–39.

27. Ibid., 729–30.

28. Peter Heenan, "Testimony to the Legislative Committee on the Department of Lands and Forests" (1941), 739.

29. Lambert and Pross, *Renewing Nature's Wealth*, 29

30. Zavitz toured the Quebec Ranger School as part of a commonwealth forestry conference. Another of his contributions to commonwealth forestry during the Second World War was to study insect problems that were plaguing the native cedars of Bermuda.

    The Dorset facility was first started under Zavitz's direction when he headed the Forest Protection Branch in the 1920s as a school to train his forest rangers, but it later took on more comprehensive training purposes. Later known as the Leslie Frost Centre, the facility's role was later expanded to include functions of hiking trails, ski trails, and a children's summer camp; it continued to be used as a training facility by the Minstry of Natural Resources until its controversial closing in 2004. There was considerable concern that the facility would be sold for a condominium development. In order to prevent this, a non-profit group was formed, the Friends of the Leslie Frost Centre, which made a bid to lease the property from the Ontario government for environmental training purposes. Another group made a bid, the Frost Centre Institute, whose bid was accepted. It was spearheaded by a retired IBM Executive, A.L Auby. In addition to running a summer camp, it developed a partnership with the University of Guelph to study vernal (temporary forested) pools. However, it folded in 2010, and the Friends of the Leslie Frost Centre are working on a new proposal.

31. Richardson, *Conservation for the People*, 3–12.

— Notes —

32. *Ibid.*, 13–14.

33. A.H. Richardson, *A Report on the Ganaraska Watershed*, published jointly by the Dominion and Provincial Governments (1944), 1–20.

34. *Ibid.*, 4–20; Richardson, *Conservation by the People*, 15, 16.

35. Ganaraska Conservation Authority, *see* "Ganaraska Forest" www.grca.on.

36.. Edmund Zavitz, "Reforestation as a Means of Controlling Runoff," in *River Valley Development in Southern Ontario*, Papers and Proceedings of the Conference on River Valley Development in Southern Ontario Held in London, Ontario, October 13th and 14th, 1944–1945, 60.

36. *Ibid.*, 59.

37. *Ibid.*

38. *Ibid.*

39. *Ibid.*, 60.

40. Edmund Zavitz, *History of Reforestation in Ontario*, (Toronto: Department of Lands and Forests, 1961), 22.

41. Zavitz, "Brief and Testimony to the Royal Commission on Timber," AO, RG 19-125, B 2495319.

42. *Ibid.*, Harold Zavitz, "Brief and Testimony to the Royal Commission on Timber."

43. *Ibid.*, Monroe Landon, "Brief and Testimony to the Royal Commission on Timber."

44. *Ibid.*, "Brief of Ontario Federation of Anglers and Hunters to Royal Commission on Timber."

45. *Ibid.*

Chapter 9 — Implementing the Vision

1. Ken Armson, *Ontario Forests: A Historical Perspective* (Toronto: Fitzhenry & Whiteside, 2001), 122. The environmental commissioner for Ontario has said that the goal is 30 percent, although some experts, notably fishery biologists, recommended a higher figure of 50 percent. The reason that another billion trees are required is that the goal of 30 percent should be throughout the whole of Southern Ontario. This means that *all* the historic county divisions and major rural watersheds in Southern Ontario would have 30 percent forest cover in their rural landscapes. This is a big challenge. While all of Eastern Ontario has at least 20 percent cover, many of the counties of the southwest have less than 10 percent.

2. A.H. Richardson, *Conservation by the People*, 22.

3. Ibid., 28.

4. Ibid., 55; Bruce Mitchell and Dan Shrubsale, *Ontario Conservation Authorities: Myth and Reality* (Waterloo: Department of Geography: University of Waterloo, 1992), 86; Ganaraska Conservation Authority, www.ganaraskagrca.on.

5. "Bruce County Forest," Bruce County, www.brucecounty.on.

6. Edmund Zavitz, "Reforestation in Ontario." Reprinted and expanded as a booklet by the Department of Lands and Forests, np. University of Toronto Archives, Collection of Publications of Edmund Zavitz. Originally published by *Canadian Geographic Journal* (April 1947), 1–29.

7. Harry Barrett, "*They Had A Dream*," 108.

8. Ibid., 9.

9. I.C. Marritt, "Beaver Released in Halton County," *Sylva*, Vol.3, No.5 (November 3, 1950). Every issue of *Sylva*, published by the

Department of Lands and Forests, can be found at St. Williams Forestry Interpretive Centre Library.

10. Lambert and Pross, *Renewing Nature's Wealth*, 436–37.

11. *Ibid.*, 434.

12. Lakehead Conservation Authority, www.lakeheadca.com.

13. G.E. Meyer, "River Shore Reservations for Kapuskasing Brook Trout," *Sylva*, Vol. 4. (1951).

14. Armson, *Ontario Forests*, 154.

15. *Ibid.*, 154.

16. Management Plan for the Limerick Forest, United Counties of Leeds and Greenville, www.uclg.ca.

17. Management Plan for the Marlborough Forest, City of Ottawa, www.ottawa.ca.

18. Ottawa Field Naturalists Club, "Ottawa," in W.W. Judd and Murray Speirs, eds., *A Naturalists Guide to Ontario* (Toronto: University of Toronto Press, 1964), 138–42.

19, Zavitz, "Recollections," 26.

20. Management Plan for Renfrew County Forest, Renfrew County, www.countyofrenfrew.on.

21. Moira Conservation Authority, "History of the Moira Valley Conservation Authority" (1958), *passim*; Department of Planning and Development, "Moira Conservation Report" (1955), *passim*.

22. Zavitz, "Recollections," 26.

23. Barrett, *They Had a Dream*, 142–43. The plaque also errs in stating Zavitz served as chief forester until his retirement in 1953 — the position had been abolished earlier. These errors that magnify the

height of Zavitz's government positions serve to highlight the great difficulties he faced in making his conservationist achievements.

24. Zavitz, "Recollections," 26; William Botti and Michael D. Moore, *Michigan's State Forests: A Century of Stewardship* (East Lansing, MI: Michigan State University Press, 2004), 22–32.

25. Zavitz, "Recollections," 26.

26. Brenlee Robinson, "From Blowsands to Mixed Forest" (2005), 25–29.

27. "Don Valley Conservation Report" (Toronto: Department of Planning and Development (1951), *passim*.

28. "Central Lake Ontario Conservation Report" (Toronto: Department of Planning and Development, 1960), Section Three Forest, 1–13; Central Lake Ontario Conservation Authority, www.cloca.com.

29. Essex Conservation Authority, www.erca.org.

30. Upper Thames Conservation Authority, www.thamesriver.on.

31. "Conservation Areas," Lower Thames Conservation Authority, www.lowerthames-conservation.on.

32. Richardson, *Conservation for the People*, 56–57; "Ausable Bayfield Conservation Authority: Watershed Report Card 2007"; Ausable Bayfield Conservation Authority, www.abca.on.

33. Conservation Authorities Branch, Department of Lands and Forests, 1963, "Big Creek Region Conservation Report, 1963," 82, 84.

34. Barrett, *They Had a Dream*, 105–06.

35. Harold Zavitz, *The History of the Lake Erie Forest District* (Toronto: Department of Lands and Forests, 1963), 25–30.

— *Notes* —

36. *Ibid., passim.*

37. Long Point Conservation Authority, www.lprca.on.

38. Harold Zavitz, *The History of the Lake Erie Forest District, passim.*

39. Edmund Zavitz, *Hardwood Trees with Bark Characteristics* (Toronto: Ministry of Natural Resources, 1973), *passim.*

40. John Robarts, "Foreword," in Lambert and Pross, *Renewing Nature's Wealth*, xiii–xiv.

41. Barrett, *They Had a Dream*, 104.

Bibliography

PRIMARY SOURCES

Archival Collections

Archives of Ontario
  E.C. Drury Papers

Fort Erie Museum
  Files on Edmund Zavitz and Zavitz Family

University of Toronto Archives
  Collection of Publications and Resume of Edmund Zavitz
  J.H. White Papers
  John Squair Papers

Government Publications

Clark, Judson. "Report of the Chief Forester of Ontario." Ontario Sessional Papers, 1904.

———. "Report," August 22, 1922, Annual Report of the Department of Lands and Forests, Ontario Sessional Papers, Ontario Legislature (1922–23).

Larose, Ferdinand. "The South Nation and Its Environs: Conservation in Eastern Ontario." Toronto: King's Printer, 1947.

Reports of the Experimental Union, 1900–1904, Ontario Sessional Papers, Ontario Department of Agriculture.

Richardson, A.H. "A Report on the Ganaraska Watershed." Toronto: King's Printer, 1944.

Interviews

Ken Armson (retired chief forester of Ontario), July 16, 2008, in St. Williams, and December 11, 2010, in Toronto.

Ed Borczon (author, community advisor, Trees Canada), Toronto, February 2010 to April 2011.

Robert James (geologist, author), Niagara-on-the-Lake, July 23, 2009.

Don Pearson (executive director, Conservation Ontario), February 4, 2011, at a forestry conference in Alliston, Ontario.

Kathleeen Mackenzie (granddaughter), Kelowna, British Columbia, January 22, 2011.

Brian Swaile (retired manager, Ontario Seed Tree Plant), February 3, 2010, in Alliston, Ontario.

Brenlee Robinson (vice-president, Ontario Urban Forest Council) Toronto, December 14, 2010.

Mel Swart (now deceased, retired member Ontario legislature for Thorold), February 20, 2005.

Dolf Wynia (president, St. Williams Forestry Interpretive Centre), January 2010 to March 2011.

Peter Zavitz (grandson), Georgetown, January 29, 2011 and April 10, 2011.

— *Bibliography* —

Management Plans for County Forests

Many of the Management Plans for County Forests contain important historical information. They are available on municipal websites.
- Management Plan for Dufferin Forest, Dufferin County
- Management Plan for Marlborough Forest, City of Ottawa
- Management Plan for Waterloo Forest Region of Waterloo
- Management Plan for York Forest, Region of York
- Management Plan for Limerick Forest, United Counties of Leeds and Grenville

Reports of Commission of Conservation

C.D. Howe, B.E. Fernow, and J.H. White. "Trent Watershed Survey." Toronto: Bryant Press, 1913.

Clyde Leavitt. "Report on the Commission on Forests, Eighth General Meeting, Commission of Conservation." Montreal: Federated Press, 1919.

Reports of the Conservation Authorities

Those consulted for this biography:
- "South Nation Valley Interim Report, 1947."
- "Don Valley Conservation Report, 1951."
- "Central Lake Ontario Conservation Report, 1961."
- "Big Creek Region Conservation Report, 1963."

Sylva

From 1948 to 1961, the Department of Lands and Forest published an excellent magazine, *Sylva*. I was able to read through the entire run of this publication. The articles in *Sylva* are illustrative of the conservationist strength of the department shaped by Edmund Zavitz, following its reorganization in 1941.

SECONDARY SOURCES

Articles

Armson, Ken, W. Ross Grinnel, and Fred Robinson, "History of Reforestation in Ontario." In Robert G. Wagner and Stephen Colombo, eds. *Regenerating the Canadian Forest*. Markham, ON: Fitzhenry & Whiteside, 2011.

Baker, Richard St. Barbe. "Baker Remembers Men of the Trees-Diamond Jubilee." In *Trees: Sixty Years Towards the Future*. Crawley Down, UK: The International Society for Planting and Protection of Trees, 1983.

Buttle, J.M. "Hydrological Response to Reforestation in the Ganaraska River Basin in Southern Ontario." www.interscience.wiley.com/journal.

Carter, John C. "Ontario Conservation Authorities: Their Heritage Resources and Museums." *Ontario History*, Vol. 88, No. 2 (Spring 2002).

Kelly, Kenneth. "Damaged and Efficient Landscapes in Rural Southern Ontario." *Ontario History*, Vol. 67, No. 1 (March, 1974).

Kuhlberg, Mark. "Ontario's Nascent Environmentalists: Seeing the Forests for the Trees in Southern Ontario, 1909–29." *Ontario History*, Vol. 94, No. 1 (June 1996).

Pearce, Bruce M. "Norfolk's Reforestation Story: The St. Williams Forestry Station." In *Historic Highlights of Norfolk County*. Hamilton, ON: Giffen and Richmond, 1979.

Pike, Robert. "Hell and High Water." *American Heritage Magazine*, Vol. 18, No. 2 (February 1967).

Pross, Paul. "The Development of Professions in the Public Service: The Foresters of Ontario." In *Canadian Public Administration* Vol. 10, No.1 (1967).

Stuart, Ian. "John Dryden." In *Canadian Dictionary of Biography*, 1901–1911, Volume XIII. Toronto: University of Toronto Press, 1994.

Weaver, Sally. "The Iroquois: the Consolidation of the Grand River Reserve in the Mid-Nineteenth Century." In Edward Rogers and Donald Smith, eds. *Aboriginal Ontario*. Toronto: Dundurn Press, 1994.

Williams, Mack. "Editor's Report," *Southern Ontario Newsletter* of the Canadian Institute of Forestry (Spring 2006).

Books

Armson, Ken. *Ontario Forests*. Toronto: Fitzhenry & Whiteside, 2001.

Barnes, Michael. *Killer in the Bush: The Great Fires of Northeastern Ontario*. Eden Mills: Boston Mills Press, 1987.

Barrett, Harry B. *They Had a Dream: A History of the St. Williams Forestry Station*. Port Rowan, ON: South Walshingham Heritage Association, 2000.

Borczon, Edward. *Evergreen Challenge*. Toronto: Ministry of Natural Resources, 1982.

Blanchet, Peter. *Forest Fires: The Study of War*. Montreal: Cantos, 2005.

Coons, C.F. "Reforestation of Private Lands in Ontario." Toronto: Forestry Research Group, Armson Private Lands Forestry Review, 1981.

Dempsey, David. *Ruin and Recovery: Michigan's Rise as a Conservation Leader*. Ann Arbor, MI: University of Michigan Press, 2002.

Drury, E.C. *Farmer-Premier*. Toronto: University of Toronto Press, 1965.

Drushka, Ken. *H.R: A Biography of H. R. MacMillan*. Madeira Park, BC: Harbour Publishing, 1995.

Fort Erie Museum, *Many Voices*. Fort Erie, ON: The Fort Erie Historical Board, 2004.

Hayes, Adrian. *Parry Sound: Gateway to Northern Ontario*. Toronto: Natural Heritage Books, 2005.

Hodgins, Bruce and Bendickson, Jamie. *The Temagami Experience: Recreation, Resources and Aboriginal Rights in the Northern Ontario Wilderness*. Toronto: University of Toronto Press, 1989.

Hosie, R.C. *Forest Regeneration in Ontario*. Toronto: University of Toronto Press, 1963.

Johnson, Charles. *E.C. Drury: Agrarian Idealist*. Toronto: University of Toronto Press, 1991.

Johnston, W. Stafford and Johnston, J.M. *History of Perth County to 1967*. Stratford: Perth County, 1967.

Killam, Gerald. *Protected Places: A History of Ontario's Provincial Park System*. Toronto: Dundurn Press and Ministry of Natural Resources, 1995.

Kuhlberg, Mark. *One Hundred Rings and Counting; Forestry Education in Toronto and Canada, 1907–2007*. Toronto: University of Toronto Press, 2009.

Lambert, Richard and Paul Pross. *Renewing Nature's Wealth*. Toronto: Department of Lands and Forests, 1967.

MacGregor, Roy. *Northern Light: The enduring Mystery of Tom Thomson and the Woman Who Loved Him*. Toronto: Random House Canada, 2010.

MacKay, Donald. *Heritage Lost*. Toronto: Macmillan Canada, 1985.

McClement, Frank. *The Burning Forest*. Toronto: McClelland & Stewart, 1969.

Miller, Char. *Gifford Pinchot and the Making of Modern Environmentalism*. Washington, DC: Island Press, 2004.

Mitchell, Bruce and Dan Shrubsale. *Ontario Conservation Authorities: Myth and Reality*. Waterloo, ON: Department of Geography, University of Waterloo, 1982.

Poling, Jim Sr. *Tom Thomson: The Life and Mysterious Death of the Famous Canadian Painter*. Canmore, AB: Altitude Publishing, 2005.

Pyne, Stephen. *Awful Splendour: A Fire History of Canada*. Vancouver: University of British Columbia Press, 2006.

Rodgers, Andrew Denny. *Bernard Eduard Fernow: A Study of North American Forestry*. Princeton: Princeton University Press, 1950.

Ross, Alexander and Peter Crowley. *The College on the Hill*. Toronto: Dundurn Press, 1995.

Saunders, Audrey. *Algonquin Story*. Toronto: Ontario Department of Lands and Forests, 1963.

Sisan, Edward. *Forestry Education at the University of Toronto*. Toronto: University of Toronto Press, 1965.

Squair, John. *History of Darlington and Clarke Townships*. Toronto: University of Toronto Press, 1927.

Strickland, Dan. "Trees of Algonquin Park." Whitney, ON: Ontario Ministry of Natural Resources and the Friends of Algonquin Park, 1989.

Robinson, Brenlee, "From Blowsands to Mixed Hardwood Forest: The History of the York Region Forest." M.A. thesis, University of Toronto School of Forestry (2005).

Walker, Laurence C. *Forests: A Naturalists Guide to Trees and Forest Ecology*. Toronto: John Wiley, 1991.

West, Bruce. *Firebirds*. Toronto: Ontario Ministry of Natural Resources, 1977.

Zavitz, Harold C. *A History of the Lake Erie Forest District*. Toronto: Department of Lands and Forests, 1963.

Websites

Ausable Bayfield Conservation Authority, "Watershed Report Card, 2007," www.abca.on.ca.

Catfish Creek Conservation Authority, "Springwater Conservation Area," www.catfishcreek.ca.

Dufferin County, "Management Plan of Dufferin County Forest," www.dufferincounty.on.ca.

Essex County Conservation Authority, "2011–16 Strategic Plan," www.erca.org.

Ganaraska Conservation Authority, "Ganaraska Forest," www.grca.on.

Lanark County, "Lanark Forest Management Plan," www.countyoflanark.on.ca.

Lower Thames Conservation Authority, "Skunk's Misery Complex,"

www.lowerthames-conservation.on.ca.

Leeds and Grenville Counties, "Management Plan of Limerick Forest," www.uclg.ca.

Oak Ridges Moraine Foundation, "Oak Ridges Moraine Fact Sheets," www.ormf.com.

Ontario Parks, "Fifty Anniversary of Springwater Provincial Park," "Turkey Point," www.ontarioparks.com.

"Sandbanks Provincial Park Draft Management Plan," www.ontarioparks.com.

Ottawa City, "Management Plan of Marlborough Forest," www.ottawa.ca.

Ottawa Field Naturalist, "Bio-Blitz of Larose Forest," www.ofnc.ca.

Renfrew County, "Renfrew County Management Plan," www.countyofrenfrew.on.ca.

Richmond Hill Field Naturalists, "Save David Dunlap Observatory," www.rhnaturalists.ca.

York Region Forest Master Plan, "Smartwood Certification Organization," www.smartwood.org.

United Counties of Prescott and Russell, "Protection and Development Plan of the Larose Forest," www.Prescott-Russell-on.ca

Index

Agreement Forest Program, 26, 45, 85, 129, 131, 132, 151, 153–56, 159, 161, 182, 191, 204, 206–09, 214, 222, 223
Air pollution, 118, 174
Aircraft, use in forest protection, 105, 142, 143, 146, 166
Alfred Bog (Russell County), 158, 160
Algoma Hills, 100
Algoma Hills Forest Reserve (Algoma Reserve), 100
Algonquin Park, 53, 100–07, 126, 127, 142, 164, 167, 202, 206, 208
Algonquin Park Act (1893), 53
American Forestry Congress, 52
Anglin, Township of (Algonquin Park), 106
Angus, Ontario, 69, 114, 124, 125, 183
Angus Municipal Forest (Simcoe County), 201, 223
Angus Plains, 57, 66
Anslow, R. Philip, 92
Armson, Ken A., 11, 13, 14
Aspen Poplar (Little Clay Belt), 71
Atcheson, Keith, 175
Athol, community of, 133
Atlantic Salmon, 156
Ausable River, 198
Ausable River Conservation Authority, 198

Backus Woods (Norfolk County), 41, 216

Baker, Richard St. Barbe, 134, 135, 171, 181–83
Balls Falls Conservation Area, 42
Bancroft, Ontario, 54
Barnes, Al, 173
Barrett, Harry B., 85, 150
Barrie, Ontario, 44, 66, 123, 125, 129, 178
Bartlett, G.W., 104
Bayly, G.H.U. "Terk," 193
Beadle, Delos "D.W.," 52, 56, 119
Beall, Thomas, 52
Bear Island (Temagami), 143
Beaver, 126, 159–61, 167, 191, 200, 202, 204
    reintroduction to Southern Ontario, 160, 202
Bedard, Joseph A., 89, 209, 211
Beeton, Ontario, 79, 129, 151
Belleville, Ontario, 33
Bertie Township (Welland County), 29, 30, 35, 37, 45, 46, 221
Bertie Township Historical Society, 46
Bertie Township School, No. 11, 35
Big Bend Conservation Area (Middlesex County), 213
Big Creek Conservation Authority (see also Long Point Region Conservation Authority), 213, 215, 216
Biltmore Estate, North Carolina, 27
Bishop, Township of (Algonquin Park), 106

Black, Robin, 137
Black Bears, 67, 160
Black Creek Pioneer Village (Toronto), 24
Black Spruce, 204
Blue Mer Bog (Ottawa Greenbelt), 207–08
Borczon, Brandon, 131
Borczon, Ed, 13
Bourget Desert (Russell County), 156, 158
Bowman, Beniah "Ben," 136, 137, 140, 141, 188
Bowmanville, Ontario, 156
Bracken, John, 40, 41, 87
Brampton, Ontario, 88, 216
Brant, County of, 120
Briggs, Frank Campbell, 120
Brighton, Ontario, 151
British Columbia, Province of, 65, 88, 137, 141
British Columbia Forest Service, 88
British Empire, 89, 91, 134
British Empire Forestry Conference (see also Imperial Forestry Conference), 9, 143
Brodie, J.A., 175
Brodie, William, 105
Brook Trout, 156, 191, 200, 202, 205, 214
    as indicator species, 205, 214
Brown, James, 37
Brown, William, 37, 40, 52, 56
Brown's Woods (OAC/University of Guelph), 38
Bruce, County of, 100

Bruce County Forest, 200
Bruce Trail, 182
Brunelle, Honourable René, 160, 218
Buffalo, New York, 36, 39

Cain, Walter, 137, 175, 188, 210
Camp Borden, 124, 129, 174
Campbell, R.H., 90
Canadian Federation of Lip Reading Clubs, 40
Canadian Forest Service, 41, 57, 88, 90, 107, 134
Canadian Forestry Association, 14, 41, 62, 63, 137, 209, 224
Canadian Geographic, 201, 236
Canadian Institute of Forest Engineers (later Canadian Institute of Forestry), 189
Canadian Institute of Professional Foresters, 189
Canadian Pacific Railway (CPR), 67
Canadian Peace and Arbitration Society, 37
Canadian Shield, 22, 50, 51, 53, 56, 86, 87, 90, 99, 100, 102, 107, 118, 161, 162, 164, 181, 208, 215, 233
Carleton, County of, 206-07
Carleton, Sir Guy (Lord Dorchester), 147
Carolinian Canada, 49, 86, 122, 148, 216
Carolinian Forest, 28, 41, 70, 85, 132, 148
Catfish Creek Conservation Authority (Elgin County), 214, 264
Catholic Church (Quebec), 54, 89, 91, 158, 209
Central Experimental Farm (Ottawa), 232
Central Lake Ontario Conservation Authority, 212, 259
Charlottesville, Township of (Norfolk County), 83
Charlotteville, Garrison and Naval Arsenal (Lake Erie), 147
Charolttenburg, Township of (Carleton County), 206
Chinquapin Oak, 148
Choquette, Monseigneur Charles P., 89
Clark, Judson, 38, 41-43, 56, 63, 65, 67, 72, 83, 91, 97, 99, 100, 137, 138, 202, 218, 221
Clarke, Irwin & Company, 175
Clarke, Township of (Durham County), 32, 81, 155

Clay Belt, 71, 72, 93, 100, 104, 105, 111, 113, 117, 135, 136, 203, 234
Little Clay Belt, 139, 140
Cobourg Creek, 156
Cochrane, District of, 144, 245
Cochrane, Francis "Frank," 64, 65, 67, 71, 72, 83, 84, 90, 136, 140, 234
Cochrane, Ontario, 65, 89, 111
Coldwater, Ontario, 37, 151
Collingwood, Ontario, 200
Commission of Conservation (Ottawa), 54, 55, 89, 90, 99, 107, 115, 136, 165, 189, 232–34, 259
Forestry Committee, 90
Commission on Agriculture (Ontario), 56
Conant, Gordon, 185
Conservation Authorities Act (1946), 26
Conservation Authorities Branch (Ontario), 173
Conservative Party, 25, 46, 57, 65, 82, 91, 95, 100, 136, 139, 141, 145, 150, 175, 186, 191, 223
Coons, Clarence F., 14, 92
Co-operative Commonwealth Federation (CCF), 175, 189
League for Social Reconstruction, 189
Co-operative Planting Program, 74, 97, 122, 193, 194, 222
Cornell University, 186
Coronation Park (Toronto), 184, 249
Counties Reforestation Act (1911), 123, 222
Craig, Roland, 57–61
Cranston, A., 44
Creelman, George, 98, 99, 238
Crosbie, Lester H., 165
Crown, H., 58
Crown Land(s), 22, 53, 54, 59, 64, 66, 72, 102, 103, 115, 135, 137, 138, 164–68, 188, 195, 202–05, 207, 224
Crown Lands Divison, 53, 59
Crown Timber Agents, 72, 91, 107, 113, 137, 138, 165, 168, 169, 174, 176, 188, 195
Crow's Pass Conservation Area (Oak Ridges Moraine), 212
Crystal Beach (Lake Erie), 36, 38
Cumberland, Township of (Carleton County), 180, 206
Cutler, Eber, 38, 39, 73, 80, 216, 221

Cutler's Dry Goods (Ridgeway), 39

Dallyn, Gordon, 201
Dance, Township of (Fort Francis District), 174
Dance Forest Fire, 110
Darlington, Township of (Durham County/Regional Municipality of Durham), 33–35, 128, 155
David Dunlap Observatory (Richmond Hill), 183, 249
Davies, Owen, 128
Davis, Graeme, 15
Davis, Premier William, 24, 46
de Casson, François Dollier, 85
de Galinée, Father René de Bréhant, 85
Delaney, Wallace A., 211
Demonstration Forest Program, 129, 162
Demonstration Woodlot Program, 150, 161
Department of Agriculture (Federal), 119
 Experimental Farms Branch, 51
 St. Catharines Research Station, 119
Department of Agriculture (Ontario), 52, 62, 65, 80, 97, 157
Department of Fish and Game (Ontario), 122, 195
Department of Lands and Forests (Ontario), 17, 23, 45, 73, 93, 97, 122, 127, 129, 137, 144, 168, 174, 186, 188, 189, 193, 195, 200, 201, 203–05, 207, 210, 216, 217, 223
 Department of Forests, 138, 145, 162–66, 172, 173, 174, 223
 Forest Protection Branch, 54, 91, 107, 118, 135, 142, 145, 146, 161, 162, 222, 250
 Kirkwood District, 163, 203
 Lake Erie District, 122, 215, 216
 Timber Administration, 137
 Tweed District, 161
Department of Lands, Forests and Mines (Ontario), 91 93, 97, 114
 Northern Development Branch, 112

— Index —

Woods and Forest Branch, later Timber Management Branch, 107, 113, 137
Department of Mines (Ontario), 135
Department of Planning and Development (Ontario), 212
Depression, *see* Great Depression
Diefenbaker, John George, 134, 181
Doherty, Manning, 120, 122, 126
Dome Gold Mine (Timmins), 89
Don River, 212
Don River Watershed Plan, 212
Dorchester, Lord, *see* Carleton, Sir Guy (Lord Dorchester)
Dorset, Ontario, 189, 250
Douglas Fir, 88
Drew, Premier George, 186, 191, 206, 215
Drury, Jim, 218
Drury, Premier Ernest Charles "E.C.," 14, 25, 44, 45, 57–62, 66, 67, 69, 86, 95, 97, 99, 116, 119, 121–30, 132, 134–41, 145, 151, 156, 161, 169, 177, 178, 180, 218, 220, 222, 223
Dryden,
  Elizabeth, 40, 44
  Jessie, *see* Zavitz, Jessie
  John, 25, 40, 44, 45, 55, 56, 58–61, 62, 65, 67
Dufferin, County of, 15, 153, 154
Dufferin County Forest, 14, 259
Dundas, County of, 128, 157, 190
Dunnville, Ontario, 217
Durham, County of, 81
Durham County Forest, 14, 190
Durham, Region of, 40, 81, 185
Durham Regional Forest, 152

Eastern Forest Reserve (Frontenac County), 54, 164, 165
Eastern Provincial Forest (formerly Eastern Forest Reserve), 165
Edmund Zavitz Forest (Norfolk County), 26, 225
Edwards, Senator William C., 88, 158, 180
Ekrid Forest (Middlesex County), 213
Elizabethtown Township (United County of Leeds and Grenville), 128

Ellice Swamp (Perth County), 213
Elliott Lake, Ontario, 100
Enniskillen Conservation Area (Oak Ridges Moraine), 212
Essex, County of, 24, 194, 213
Essex County Conservation Authority, 213
Etobicoke Creek, 198
Etobicoke Creek Conservation Authority (now Toronto and Region Conservation Authority), 198
Experimental Farms Branch, *see* Department of Agriculture (Federal)
Experimental Union (OAC Alumni), 37, 45, 56–59, 61, 74, 87, 98, 99

"Farm Forestry," 79
Farmer's Advocate, The, 52, 177
Farmers' Institutes, 10, 66, 95
Faull, Dr. J.H., 59, 60, 118
Federation of Ontario Naturalists, 126, 182
Ferguson, Premier Howard, 25, 95, 112, 113, 116, 141, 143–45, 147, 150, 223
Fernow, Bernard Eduard "B.E." (Dean), 52, 70, 71, 73, 136, 186, 218, 234
Field Day(s) (OCRA), 178–80, 185, 190
Fifty Years of Reforestation in Ontario, 201, 217, 224
Finlayson, Ernest, 134
Finlayson, William, 73
First World War, 17, 37, 74, 97, 98, 102, 109, 118, 124, 125, 167, 182, 184
Fletcher, James, 51
Forest cover, 19, 22, 24, 27, 29, 30, 49, 61, 98, 99, 123, 157, 158, 160, 177, 183, 191, 193–95, 197–203, 208, 213–15, 218, 224, 252
Forest cover, percentage of,
  in Eastern Ontario, 38%, 24
  in Essex County, 5%, 24
  in Ganaraska Watershed, 43.6%, 198
  in Russell County, 4%, 158
  in Southern Ontario, 25.2%, 197
  in Southwestern Ontario, 17%, 218
  Upper Thames Conservation Authority, 11.5%, 213

watershed protection goals, 50%, 49
Forest Fire Prevention Act (1917), 113–18, 138, 140, 141, 194
Forest fires, 17, 19, 21, 22, 24, 45, 50, 57, 63, 67, 88–91, 97, 99, 104, 105–12, 115–19, 138, 139, 141, 144, 146, 158, 162, 171, 174, 176, 194, 220, 222, 233
Forest Protection Branch, *see* Department of Lands and Forests
Forest Reserve, 54, 56, 59, 65, 100, 104, 164, 165, 202, 204, 223
Forestry Faculty, *see* University of Toronto
Forestry Station #1, *see* St. Williams Forestry Station
Forestry Station #2, *see* Turkey Point
Forestville, Ontario, 31, 169, 195, 216, 220
Fort Erie, Ontario, 29, 46, 49, 221
Fort Francis, District of, 110, 174
Fort Norfolk (Lake Erie), 147
Fort William, Ontario, 136, 203
Fort William Reforestation Station, 204
Fowlds, Walter, 155
Franklin Island Provincial Park (Georgian Bay), 126
French River, 100, 127, 135, 145
Freswick, Township of (Algonquin Park), 106
Frontenac, County of, 162
Fruit Growers of Ontario, *see* Ontario Fruit Growers Association

Gads Hill Swamp (Perth County), 213
Gamey, Robert Roswell, 136, 243
Ganaraska Conservation Authority, 197–99
Ganaraska Forest, 25, 197, 198, 200
Ganaraska River, 21, 22, 32, 190, 191, 193, 194, 199, 201, 214, 227
Ganaraska Watershed, 23, 25, 31, 190, 191, 197, 212
Ganaraska Watershed Survey, *see* "Report on the Ganaraska Watershed, A"

269

Georgian Bay Provincial Forest, 164, 166
Gibson, John Allen, 211
Gillies Lumber Limit, 164
Glackmeyer Report (1960), 203
Glengarry, County of, 157
Grand River, 49, 178, 188, 193
Grand Trunk Railway, 104, 108
Grant, Village of, 53
Great Depression, 156, 171, 176, 178, 201, 206, 213, 219
Greenbelt,
  Ontario, 25, 30, 193, 200
  Ottawa, 24, 157, 158, 207, 208
Grey, County of, 172, 173
Grey County Forest, 182
Grigg, Albert, 95, 96, 137, 238
Guelph, Ontario, 38, 40, 55, 63, 66, 69, 72, 73, 79, 91, 189, 222
Guelph Arboretum (OAC/University of Guelph), 28, 38, 79
Guelph Reforestation Station, 84

Haggerty, Ray, 46
Haileybury, Ontario, 139, 140, 141, 174
  fire of 1922, 110, 122, 140, 142
Haldimand Tract (Northumberland County), 81
Halton, County of, 202
Halton, Region of, 201
Hamilton, Ontario, 128, 230
Hanover, Ontario, 79
Hardwood Trees of Ontario: With Bark Characteristics, 217, 224
Harris, Mike, 26
Hastings, County of, 54, 86, 161, 164
Headquarters Tract (York Region) (formerly Vivian Forest), 131
Hearst, Premier William H., 90, 91, 100, 112
Heenan, Peter, 174, 176, 188, 249
Henderson, James, 153
Hendrie Tract Forest (Simcoe County), 15, 130, 131, 153, 178, 223
Henry, Premier George, 145, 150, 223
Hepburn, Premier Mitchell "Mitch," 18, 126, 145, 168, 169, 172, 175–77, 186, 188, 202, 204, 205, 209, 223
Highland Grove of White Pines (Algonquin Park), 71, 101

Hillcrest Forestville Cemetery, 220, 224
Hipel, Norman, 188
History of Darlington and Clarke Townships, The, 32
Hollinger Gold Mine (Timmins Area), 89
Hosie, R.C., 106, 262
Howard Ferguson Reforestation Station (*see also* United Counties of Leeds and Grenville), 206
Howe, Clifton Durant "C.D." (Dean), 73, 74, 86, 146
Huron, County of, 194

Imperial Forestry Conference (1928) (*see also* British Empire Forestry Conference), 133
Imperial India, 23, 27, 52, 54, 56, 114, 168, 184
Independent Labour Party, 25, 119, 136
Inglis Falls Municipal Forest (Grey County), 151
Innis, Harold, 50
International Tree Foundation (formerly Men of the Trees), 134
Iroquois Falls, Ontario, 112
Iroquois Ridge (Oak Ridges Moraine), 193
Irwin, John C.W., 175, 183, 186–88, 249

J.H. White Forest (Norfolk County), 26, 201
Jackson, Edith, 46
James Bay Lowlands, 219
Jenkins, Ted, 200, 210
Johnson, E. Pauline, 49
Johnson, George, 49
Johnston, Reverend Milton, 219
Jones, Walter, 87, 185

Kaladar, Township of (United Counties of Lennox and Addington), 209
Kapuskasing, Ontario, 203, 205
Kawartha Provincial Forest (Peterborough County), 164–66
Kemptville, Ontario, 126, 141
Kemptville Agricultural College, 126
Kennedy, A.J., 141, 146, 172
Kennedy, Major General Howard, 194–95
Kent, County of, 194

Kenya, 134, 135
Kilman, Alva, 35, 36, 44
Kilman, Leroy, 36
King, Township of, 182
King, William Lyon Mackenzie, 183
Kirkwood, Alexander, 53, 54
Kirkwood, District of, *see* Department of Lands and Forests
Kirkwood, Township of, 164
Kirkwood Desert (District of Algoma), 163
Kirkwood Forest Management Unit, 164
Koch, Henry, 28

Laflamme, Monseigneur J.C.K., 89
Lake Erie, 17, 36 53, 70, 85, 146–48, 217
Lake Erie District, *see* Department of Lands and Forests
Lake Huron, 81, 198, 214, 215
Lake Ontario, 25, 69, 97, 126, 133, 191
Lake Superior, 54, 144
Lakehead Region Conservation Authority, 204
Lambert, Richard, 23, 24
Lambton, County of, 81, 194, 216
Lanark, County of, 181–82
Landon, Ken, 215
Landon, Monroe, 18, 85, 123, 149, 177, 178, 182, 189, 194, 215, 229
Lane, George, 73
Lane, George Ritchie, 73
Larose, Ferdinand, 156, 157–61, 206, 217, 223
Larose, Mrs. Ferdinand, 160
Larose Forest (*see also* United Counties of Prescott and Russell), 53, 157, 159–61, 180, 206–08
Laurentide Air, 142
Laurier, Sir Wilfrid, 54, 183, 184, 232
Lavrielle Lake (Algonquin Park), 103
League for Social Reconstruction, *see* Co-operative Commonwealth Federation
League of Catholic Farmers, 157
Leavitt, Clyde, 99, 115, 237, 238
Leeds, County of (*see also* United Counties of Leeds and Grenville), 128, 182

— Index —

Lefebvre, Father Joseph Daniel, 28
Leslie, George, 56
Leslie Frost Centre, 250
Liberal Party, 40, 46, 65, 145, 168, 169, 172–75, 177, 183, 185, 186–89
Limerick Forest (United Counties of Leeds and Greenville), 182, 259
Lincoln, County of, 42, 67, 221
Linton, George M., 125, 154–56
Little, William, 53, 63
Little Clay Belt, 139, 140
London, Ontario, 21, 37, 51, 177, 179, 192, 193, 198, 213, 219, 251
London City Council, 219
Long Point, 146, 216, 217
Long Point Provincial Park, 126
Long Point Region Conservation Authority (Norfolk County), 216
Long Sault Conservation Authority, 212
Longstaff Prison Farm (York Region), 151
Lovejoy, P.L., 211
Lower, Arthur, 50
Lower Thames Conservation Authority, 178, 213
Lyons, Robert W., 211

MacDougall, Frank Archibald, 145, 166, 167, 188, 216, 223
Mach, Caroline, 15
Mackenzie, Kathleen, 14, 46, 186
MacMillan, H.R., 88, 209, 211
MacMillan Provincial Park, 88
MacMillan-Blodel Corporation, 88
MacNachton, Colonel, 155
Madawaska River Watershed, 127
Madden, J.F.S., 185
Madden, Margaret Irene (Henderson), *see* Zavitz, Margaret
Madoc, Township of (Hastings County), 209
Magladery, Thomas, 139, 141
Manitoba, Province of, 41, 87
Mapledorm, Clare E., 209–10
Marathon Logging Company, 166
Markham, Ontario, 26, 151
Marlborough, Township of (Carleton County), 207
Marlborough Forest, 207

Marritt, Isaac. C., 125, 150, 161
Marsh, Leonard, 189
Martin, John, 146, 147, 150
Mason, T.H., 58, 60
Matheson, Ontario, 95, 112, 113
  fire of 1916, 95, 104, 111, 222, 237
Maxville, Ontario, 157
Maxwell, Roy, 142, 223
McAmish, John M., 51
McCall, Bruce, 147
McCall, Walter, 82
McCall Furniture Factory (St. Williams), 82
McEwen, Peter, 162–64, 166, 174
McMaster University (Hamilton), 209, 224
McMaster University (Toronto), 29, 40, 41, 44, 55, 56, 130, 221, 230
McQuesten, Thomas, 128
Men of the Trees, 134, 135,
  Canadian chapter, 182–85, 189
Merrickville, Ontario, 97
Metz, Michigan, 63
  fire of 1908, 63
Middlesex, County of, 177, 178, 213, 216
Middlesex County Forest, 178
Midhurst Forestry Station, 67, 73, 129, 130, 150, 151, 153, 222, 223
Midhurst, Ontario, 15, 57, 66, 69, 125, 130
Midhurst Park (Simcoe County), 129
Midland, Ontario, 79
Miller, Bert, 45
Miller, Char, 27
Miller, Gordon, 218
Mills, Henry, 136
Mills Block Conservation Area, 204
Ministry of Natural Resources (MNR), 211
Mohawk First Nation, 27, 28
Moira River Watershed, 181, 208, 209
Moira Valley Conservation Authority, 208
Mono Cliffs Provincial Park (Dufferin County), 154
Mono Tract (Dufferin County), 154
Monteith, Ontario, 111
Monteith, Nelson, 57–59, 66, 67, 72, 78, 83, 178, 213, 219
Montgomery, Dr. Fred, 85

Moodie, Susanna, 32–33
Moose Creek Bog (Larose Forest), 159
Mosa-Bradshaw Forest (Skunk's Misery) (Middlesex County), 178, 213
Muldrew, Dr. W.H., 59, 60
Mulmur, Township of (Dufferin County), 154
Mulock, Sir William, 183–85
Municipal Parks Commissions, 127
  Hamilton, 128
  St. Catharines, 128
  Uxbridge, 128
Muskoka, District of, 50, 59, 127

Napanee Region Conservation Authority, 208
Nature Conservancy of Canada, 41, 153
New Democratic Party (NDP), 26, 46
New Liskeard, Ontario, 72, 110
New Lowell, Ontario, 129
New Zealand, 74, 134
Newman, Frank, 74, 97, 123, 148, 150
Niagara Escarpment, 19, 42, 67, 136, 151, 154, 182, 193, 200–02, 215, 219, 221
Niagara Escarpment Biosphere Reserve, 30
Niagara Escarpment Plan, 30
Niagara Peninsula Conservation Authority, 18, 30, 31, 229
Nixon, Harry, 120
Noad, Frederick, 168, 169, 172–76, 188, 202, 209, 223
Norfolk, County of, 17, 18, 24, 41, 43, 53, 61, 63, 67, 69, 82–86, 97, 123, 124, 146, 147, 150, 169, 194, 215, 216, 221, 224, 229
Norfolk 133rd Battalion, 97
Norfolk County Council, 61, 150
Northern Development Branch, *see* Department of Lands, Forests and Mines
Northumberland, County of, 81, 190
Northumberland County Forest, 155, 156
Norway Spruce, 74, 77–78

OAC Guelph Nursery, 59, 73
Oak Ridges Moraine, 19, 22, 24, 26, 31, 32, 34, 52, 56, 62, 74, 80, 81, 86, 120, 121, 125,

271

128, 131, 132, 151–56, 177, 185, 190, 191, 193, 194, 199, 200, 212, 223
Oak Ridges Moraine Foundation, 26, 153, 212, 242
Oak Ridges Moraine Trust, 153
Oakville Creek, 202
Oka, Quebec, 20, 27, 158, 228
   reforestation of, 27, 28, 158
Ontario, County of, 40, 81, 156, 185
Ontario, Province of,
   Eastern, 24, 88, 101, 128, 156–61, 179, 180, 182, 205–07, 211, 217, 223
   Northern, 21, 55, 64–66, 71, 88, 90, 93, 100, 109–12, 116, 136, 137, 144–46, 163, 167, 168, 172, 222, 224
   Southern, 9, 10, 18, 22, 28, 30, 37, 46, 50, 53, 56, 63, 66, 90, 97–99, 101, 120, 122, 123, 127, 131, 139, 144, 148, 150, 153, 156, 173, 175, 178, 179, 187, 197, 201, 202, 205, 214, 218, 223, 224
Ontario Agricultural College (OAC), 10, 35, 37, 38, 40, 55, 56, 69, 80, 136, 222, 224
Ontario Air Service, see Ontario Provincial Air Service
Ontario Archaeological and Historic Sites Board, 209, 210, 224
Ontario Conservation and Reforestation Association (OCRA), 176–78, 182
Ontario County Forest, 198
Ontario Federation of Anglers and Hunters, 189, 195
Ontario Forestry Association, 80
Ontario Fruit Growers Association, 37, 51, 52, 56
Ontario Municipal Board (OMB), 141
Ontario Paper Company (Thorold), 203
Ontario Professional Foresters Association, 73
Ontario Provincial Air Service (Forestry Branch), 142, 143, 223
Ontario Seed Tree Plant (Angus), 114, 124, 223
Ontario Woodlot Association, 80
Orangeville, Ontario, 79

Orono, Ontario, 125
Orono, Township of (Durham Region), 156
Orono Reforestation Station, 128, 154, 155, 223
Orr Lake, 57, 131
Orr Lake Forest (Simcoe County), 131, 153
Ottawa, Ontario, 51, 53, 88, 89, 91, 118, 134, 156, 158, 206–08, 232
Ottawa Field Naturalists, 51
Ottawa River, 102, 161
Ottawa-Huron Survey, 127
Owen Sound, Ontario, 151
Owen Sound Kiwannis Club, 151

Parkway Belt Plan, 219
Parry Sound, District of, 104, 127, 162, 164
Parry Sound, Ontario, 142, 164
Partridge, Ella, 66
Payne Tract (Norfolk County), 150
Perth, County of, 78, 213
Petawawa Management Unit, 206
Petawawa River, 100, 102, 105
Phipps, Robert W., 52, 56
Piche, C.C., 89
Pinchot, Gifford, 27, 28, 41, 52, 54, 57, 99, 158, 171, 228
Pinery Provincial Park (Huron County), 81
Polar Bear Provincial Park, 219
Porcupine, Ontario (Golden City), 89
   fire of 1911, 88–90
Porquois Junction, Ontario, 89
Port Arthur, Ontario, 203
Port Credit, Ontario, 219
Port Dover, Ontario, 85, 149, 150
Port Hope, Ontario, 21, 198
Port Rowan-South Walsingham Heritage Association, 224
Porter, Dana, 198
Porter, Watson, 176–80, 182, 189, 198, 213, 219, 223
Pottsville, Ontario, 89
Pound, I.L., 39, 221
Pratt, Arthur C., 82–84, 98, 123
Prescott, County of, see United Counties of Prescott and Russell
Preston, Ontario, 188
Preservation of Agricultural Lands Society (PALS), 15, 18, 46
Prince Edward, County of, 96, 97, 125, 132

Prince Edward Island, Province of, 38, 87, 185
Prince Edward Island National Park, 185
Private Landowners Tree Seedlings Assistance Program, 204
Pross, Paul, 23–24
Prout,
   Dorothy, see Zavitz, Dorothy
   Edmund, 22, 31, 32, 35, 75, 221
   Laura, see Squair, Laura
Provincial Parks Act (1921), 126
Purcell, C.R., 189

Quaker(s), Society of Friends, 36–37
Quebec, Province of, 27, 54, 88–91, 110, 111, 115, 117, 141, 157, 158, 161, 176, 188, 222, 229, 234
Quebec City, 63, 89, 91, 189
Quebec Forest Service, 89, 90, 209
Quebec Ranger School, 189, 250
Quinte Conservation Authority, 208

Rahamni Conservation Authority (Oak Ridges Moraine), 212
Rainy River, Ontario, 112
Raisin River Conservation Authority, 206
Rathwell, Marshall, 159, 180
"Recollections, 1875–1964," 31, 38, 39, 41, 211, 217, 224
Red Pine, 10, 29, 82, 85, 102, 124, 125, 158, 159, 166, 181
Redbud (Carolinian Tree), 148
Renewing Nature's Wealth, 17, 217, 219
Renfrew, County of, 100, 127, 162, 208
Renfrew County Agreement Forest, 208
"Report on Reforestation of Wastelands in Southern Ontario (1908)," 10, 61, 69, 80–82, 86, 146, 194, 205, 208, 214, 222
"Report on the Ganaraska Watershed, A," 190, 191
Richardson, Arthur Herbert "A. H.," 24–26, 58, 61, 130, 146, 173, 182, 189, 190, 191, 212, 214
Richmond Hill, Ontario, 183

— Index —

Richmond Hill Field Naturalists, 249
Ridgeway, Village of, 29, 31, 35, 36, 38–40, 44–46, 73, 221, 230
Robarts, Premier John, 24, 218, 219
Robert Hunter Provincial Park (York Region), 26
Robertson, W.J., 39
Rock Point Provincial Park (Haldimand County), 217
Rockland Plantation (Ottawa), 15, 88, 92, 158, 180, 182
Rondeau Provincial Park (Kent County), 53, 67, 70, 71, 122, 217
Roosevelt, President Franklin D., 171
Roosevelt, President Theodore, 27, 54
Roth, Dean Filbert (Dean), 62, 63, 211
Rouge River, 26
Roy, Bishop, 89
Royal Commission on Timber (1920), 194
Royal Ontario Museum (Toronto), 36, 105
Russell, County of (*see also* United Counties of Prescott and Russell), 156–59, 161
Rutledge, Edward, 35

Sandbanks Provincial Park, 69, 96, 125, 132, 133
Sandbanks Transfer Station, 132
Sauble Beach, Ontario, 198
Sault Ste. Marie, Ontario, 100, 143, 163, 166
Saunders, Charles, 51
Saunders, William E., 51, 52, 56, 232
Schaaf, Marcus, 211
Scott, Archdeacon F.G., 185
Second World War, 134, 187, 189, 191, 195, 215, 219, 250
Select Committee on Conservation, 194
Severn River Provincial Park (Georgian Bay), 126
Sharpe, J.F., 165
Sheffield, Township of (United Counties of Lennox and Addington), 209
Shisler, Peter, 29–30
Sibley Forest Reserve, 164
Sibley Provincial Forest (now Sleeping Mountain Provincial Park), 204

Sifton, Clifford, 107
Simcoe, County of, 15, 44, 45, 57, 60, 66, 67, 73, 99, 123, 124, 129–32, 150, 153, 177–79, 201
Simcoe, Elizabeth, 50
Simcoe, Lieutenant Governor John Graves, 50, 147
Simcoe, Ontario, 123, 149
Sioux Lookout, Ontario, 142
Sisson, C.B., 136, 137, 155
Skunk's Misery, *see* Mosa-Bradshaw Forest
Sleeping Giant Provincial Park, 54, 164, 205
Smith, Dr. Richard, 41, 55
Soil erosion, 51, 158, 195
South Lorrain, Township of (Timiskaming District), 164
South Nation Conservation Authority, 206
South Nation River, 182
South Porcupine, Ontario, 89
South Walsingham, Township (Norfolk County), 86
Southworth, Thomas, 41, 52, 54, 56
Sparta, Ontario, 37
Springwater Forest Conservation Area, 214
Springwater Provincial Park (Simcoe County), 67, 69
Spruce Falls Power and Paper Company, 205
Squair,
  Francis, 32, 75, 76, 80, 190, 222
  John, 32–35, 62, 80, 128, 190, 229
  Laura (Prout), 32
St. Catharines, Ontario, 46, 52, 56
St. Catharines Collegiate, 39, 221
St. Catharines Research Station (*see also* Department of Agriculture [Federal]), 119
St. Thomas, Ontario, 79, 149
St. Williams, Ontario, 13, 18, 19, 31, 82, 84, 124, 222
St. Williams Forest, 85
St. Williams Forestry Interpretive Centre, 13, 14
St. Williams Forestry Station (Station No. 1), 10, 13, 19, 73, 74, 84, 88, 91, 97, 98, 118, 123, 124, 147–50, 163, 201, 209, 210, 215, 216, 219, 224
St. Williams Nursery and Ecology Centre (former St. Williams Station), 225

Staples, A.J., 128
Steward, John, 128
Stratford, Ontario, 213
Sulpician Journals, 86
Sulpician Order, 28, 85
Sunnydale, Township of (Simcoe County), 129
Swart, Mel, 15, 18, 30, 31, 46, 194, 203, 229

Tamblyn, John, 128
Tara, Ontario, 100
Temagami, Ontario, 54, 143, 164
Temagami Provincial Forest, 164
Temagami Forest Reserve, 164
Temagami Forest Station, 143
Thames River, 21, 177, 178
Thames River Flood of 1937, 179, 213
Thompson, R., 176
Thomson, Tom, 105, 106, 112
Thunder Bay, Ontario, 204
Tile Drainage Act (1878), 26
Timber Administration Branch, *see* Department of Lands and Forests
Timber Commission (1920), 135
Timber Management Branch, *see* Department of Lands and Forests
Toronto, Ontario, 29, 40, 41, 56, 62, 66, 107, 117, 146, 151, 168, 175, 184, 185, 211, 212, 216, 219
Toronto and Region Conservation Authority, 198, 202
Trees Act (1946), 191
Trent Watershed, 165
Trent Watershed Survey (1913), 55, 86, 107, 208
Tulip Tree (Carolinian Tree), 67, 148, 214, 220
Turkey Point, Ontario, 82, 146–47
Turkey Point Forestry Station No.2, 148, 201, 215
Turkey Point Provincial Park, 26, 201, 217
Tweed, Ontario, 86
Tweed District, *see* Department of Lands and Forests

United Counties of Leeds and Grenville, 182
United Counties of Prescott and Russell, 159
United Farmers of Ontario (UFO), 25, 99, 119, 122, 136, 145

273

University of Guelph, 28, 38
University of Laval, 89, 157
University of Michigan, 29, 36, 62, 63, 66, 99, 222, 229
University of Toronto, 32, 52, 54, 55, 59, 72, 73, 97, 106, 118, 125, 126, 136, 139, 164, 175, 183, 184, 189, 209, 222, 224
  Faculty of Forestry, 54, 134, 162, 167, 201
Upper Thames Conservation Authority, 14, 58, 213, 254
U.S. Forest Service, 27, 41, 52, 54, 57, 91, 99, 114
Uxbridge, Ontario, 81, 128
Uxbridge, Township of (Ontario County, now Durham Region), 156

Vivian Forest (York Region), 120, 131, 132, 151
Vonk, Wim, 92

Walpole Island, 49
Wanapitei Provincial Forest (District of Sudbury), 164
Wardsville, Ontario (Middlesex County), 213
Waterloo, County of, 188
Welland, County of, 18, 30, 31, 221, 229
Welland County Tree-Cutting Bylaw, 31, 229
Wentworth, County of, 120
West Lake Brick Company (Athol), 133
Whitby, Ontario, 40
White, Aubrey, 67, 71, 72, 91, 228
White, James Herbert "J.H.," 54, 55, 72, 73, 86, 90, 97, 107, 108, 114, 139, 148, 161–63, 165, 181, 190, 201, 208, 233
White, John, 219
White Pine, 17, 27–29, 43, 56, 57, 59, 62, 67, 71, 75–79, 83–85, 100–07, 117, 118, 124, 126, 127, 132, 133, 148, 158, 159, 163, 182, 185, 201, 206, 207, 214, 218, 220, 228
White Pine Blister Rust, 118, 125, 142
White's Bush (*see also* Springwater Forest Conservation Area), 214, 216
Whitchurch, Township of (Durham County), 132
Whitchurch-Stouville, Municipality of, 153
Whitney, Ontario, 142

Whitney, Sir James, 95, 100, 136
Wild Turkey, 147, 148, 154
Wilderness Act (1956), 217
Williams, J.R.M. "Mack," 28, 228
Woodland Caribou, 111, 205
Woodstock, Ontario, 29, 149
Woodstock Collegiate, 29, 40, 221
Wynia, Dolf, 13, 173, 236

Yale University, 29, 36, 41, 45, 62, 89, 153, 222
York, County of, 132, 150, 151, 154, 212
York County Reforestation Committee, 153
York, Region of, 131, 153
  Forestry Department, 153
York Regional Forest, 131, 153

Zavitz,
  Charles A., 37, 40, 45, 50, 56, 58, 60, 122
  Christian, 36
  Dean Clarence (son), 67, 169
  Dorothy (mother), 22, 31, 32, 169, 221
  Edmund Ross (son), 67, 195, 218
  George, 36
  Harold, 50, 122, 215, 216
  Henry, 36
  Jacob, 36–37
  Jessie (Dryden) (First Mrs. Edmund), 25, 40, 44, 56, 66, 67, 134, 169, 186, 220–23
  John Dryden (son), 67, 169, 223
  Joseph (father), 31, 36, 221
  Margaret Irene (Henderson) (Second Mrs. Edmund), 185, 186, 219
  Peter, 14
Zavitz General Store (Ridgeway), 31
Zavitz Pines (OAC/University of Guelph), 79

About the Author

John Bacher received his doctorate from McMaster University in 1985; his focus was on the struggle for social housing in Canada. His dissertation, *Keeping to the Marketplace: The Evolution of Canadian Housing Policy* was published in 1995. His second book, *Petrotyranny*, published by Dundurn Press in 2000, examines the negative relationship across oil, war, and dictatorship. An active conservationist with a particular interest in the protection of the rural landscape of Southern Ontario, John became a founding director of the Preservation of Agricultural Lands Society (PALS) in 1976 and is currently employed by them as a researcher. He resides in St. Catharines, Ontario.

Of Related Interest

The Last Stand
A Journey Through the Ancient Cliff-Face Forest of the Niagara Escarpment
Peter E. Kelly and Douglas W. Larson
978-1897045190
$39.95

The most ancient and least disturbed forest ecosystem in eastern North America clings to the vertical cliffs of the Niagara Escarpment. Prior to 1988 it had escaped detection even though the entire forest was in plain view and was being visited by thousands upon thousands of people every year. *The Last Stand* reveals the complete account of the discovery of this ancient forest, of the miraculous properties of the trees forming this forest (eastern white cedar), and of what it was like for researchers to live, work, and study within it. The unique story is accompanied by stunning colour photographs and through vivid first-hand accounts.

The Legacy of John Waldie and Sons
A History of the Victoria Harbour Lumber Company
Kenneth A. Armson and Marjorie McLeod
978-1550027587
$22.99

At the time of his death in 1907, John Waldie, founder of the Victoria Harbour Lumber Company, was identified as "the second largest lumber operator in Canada." Through extensive documentation of Waldie's logging practices, as well as descriptions of the forests he harvested as they are today, *The Legacy of John Waldie and Sons* provides insights into days of rampant entrepreneurialism, the world of the lumber barons, and the overall impact they had on our Ontario forests.

Alligators of the North
The Story of the West & Peachey Steam Warping Tugs
Harry B. Barrett and Clarence F. Coons
978-1554887118
$34.99

With historical photographs and thoughtful prose, *Alligators of the North* examines the role that the Alligator tug played in opening the Ottawa River to the timber industry. The Alligator tug, a little-known workhorse, was a major boon to the development of early Northern Ontario, and this text serves as a fitting tribute to its lasting impact on the region.

Available at your favourite bookseller.

DUNDURN
www.dundurn.com

What did you think of this book?
Visit *www.dundurn.com* for reviews, videos, updates, and more!

www.ingramcontent.com/pod-product-compliance
Lightning Source LLC
Chambersburg PA
CBHW071425150426
43191CB00008B/1042